15°
20°
25°
30°
35°
40°
45°
50°E
65°
60°N
55°N
50°N
45°N
40°N
35°N
30°N
15°E
20°E
25°
30°
35°
40°
45°
50°E
FINLAND
SWEDEN
Turku
Lahti
Helsinki
Leningrad
Tallinn
Stockholm
Motala
Baltic Sea
USSR
Riga
Hörby
Moscow
Kaliningrad
Kaunas
Vilnius
Minsk
POLAND
Szczecin
Stargard
Kuibyshev
Berlin
Königswusterhausen
Warsaw
Konstantinow
Dresden
Wroclaw
Liblice
Katowice
Rzeszów
Kiev
Litomysl
Lvov
CZECHOSLOVAKIA
Topolna
Vienna
Linz
Bratislava
Kosice
USSR
USTRIA
Budapest
Graz
Siofok
Szolnok
Kichinev
HUNGARY
ROMANIA
Cluj
Llubljana
Zagreb
Deanovec
Solt
Lugoj
Brasov
Tovarnik
Novi Sad
Timisoara
Belgrade
Stolnik
Bucharest
YUGOSLAVIA
Caspian Sea
Adriatic
Biograd
Sarajevo
Vidin
BULGARIA
Black Sea
Pristina
Sofia
Tbilisi
Tirana
ALBANIA
TURKEY
Erzurum
Bari
Korce
Istanbul
Kavala
Lushnje
GREECE
Etimesgut
TURKEY
Polatli
Athens
Antalaya
IRAN
Rhodes
Malta
Sea
Cyprus
Tripoli
SYRIA
Beirut
LEBANON
Damascus
Haifa
IRAQ
ISRAEL
Tel Aviv
Jerusalem
Amman
JORDAN
IBYA
EGYPT
SAUDI ARABIA
Cairo

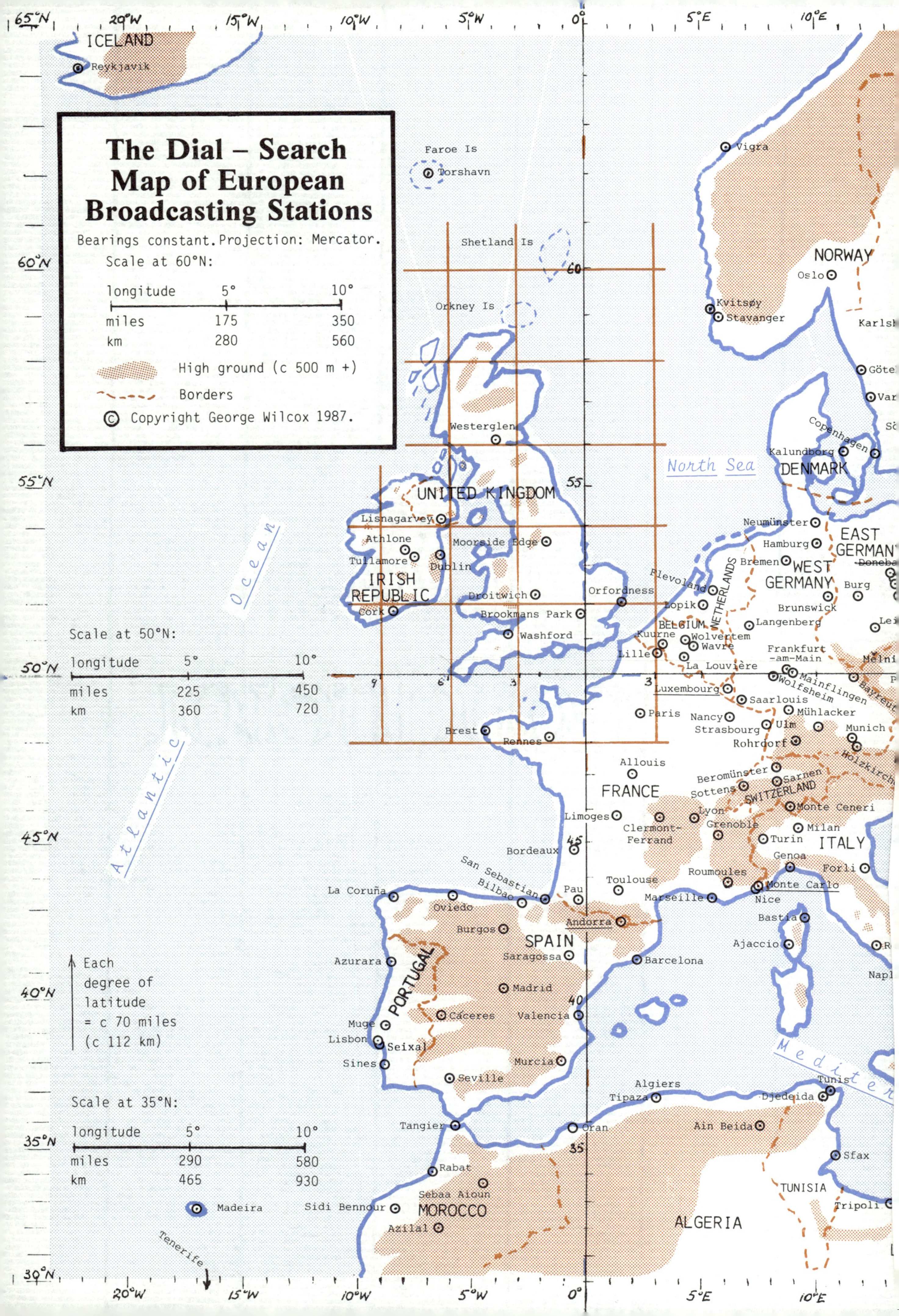
The Dial – Search Map of European Broadcasting Stations
Bearings constant. Projection: Mercator.
Scale at 60°N:
longitude 5° 10°
miles 175 350
km 280 560
High ground (c 500 m +)
Borders
© Copyright George Wilcox 1987.
Scale at 50°N:
longitude 5° 10°
miles 225 450
km 360 720
Each degree of latitude = c 70 miles (c 112 km)
Scale at 35°N:
longitude 5° 10°
miles 290 580
km 465 930
Atlantic
Ocean
North Sea
Mediter
ICELAND
Reykjavik
Faroe Is
Torshavn
Shetland Is
Orkney Is
UNITED KINGDOM
Westerglen
Lisnagarvey
Moorside Edge
Droitwich
Brookmans Park
Washford
Orfordness
IRISH REPUBLIC
Athlone
Tullamore
Dublin
Cork
NORWAY
Vigra
Oslo
Kvitsøy
Stavanger
DENMARK
Kalundborg
Copenhagen
NETHERLANDS
Flevoland
Lopik
BELGIUM
Kuurne
Wolvertem
Wavre
La Louvière
Lille
Luxembourg
WEST GERMANY
EAST GERMANY
Neumünster
Hamburg
Bremen
Burg
Brunswick
Langenberg
Frankfurt -am-Main
Mainflingen
Wolfsheim
Saarlouis
Mühlacker
Ulm
Munich
Rohrdorf
FRANCE
Paris
Nancy
Strasbourg
Brest
Rennes
Allouis
Limoges
Clermont-Ferrand
Lyon
Grenoble
Bordeaux
Pau
Toulouse
Marseille
Roumoules
Nice
Monte Carlo
SWITZERLAND
Beromünster
Sottens
Sarnen
Monte Ceneri
ITALY
Milan
Turin
Genoa
Forli
Bastia
Ajaccio
SPAIN
La Coruña
Oviedo
San Sebastian
Bilbao
Burgos
Saragossa
Andorra
Madrid
Barcelona
Valencia
Murcia
Seville
PORTUGAL
Azurara
Muge
Lisbon
Seixal
Sines
MOROCCO
Tangier
Rabat
Sebaa Aioun
Sidi Bennour
Azilal
Madeira
Tenerife
ALGERIA
Algiers
Tipaza
Oran
Ain Beida
TUNISIA
Tunis
Djedeida
Sfax
Tripoli
65°N
60°N
55°N
50°N
45°N
40°N
35°N
30°N
20°W
15°W
10°W
5°W
0°
5°E
10°E

Dial-Search

The Listener's Check-list
and Guide to
European Radio Broadcasting

Sixth Edition
1990/91

GEORGE WILCOX – EASTBOURNE

Published by George Wilcox,
9 Thurrock Close,
Eastbourne, East Sussex BN20 9NF.

First published 1981

ISBN 0 9508575 4 8

British Library Cataloguing in Publication Data

Wilcox, George
Dial-Search. — 6th ed.
1. Europe. Radio stations
I. Title
621.384094

ISBN 0-9508575-4-8

Compiled, designed and type-set
by George Wilcox.

Printed in England
by Finmere Press Ltd., Eastbourne.

CONTENTS

Finding your way in a city, you need addresses and a plan. In the country, you need a road map. Dealing with people, you need directories and phone books. All these depend on getting numbers right. The same applies to broadcasting.

Radio sets are beginning to look like calculators, with keys to press that display a number, and no knobs or dial at all. A reference list of frequencies seems essential, or where do you start? The Dial-Search checklists are your directory.

For those of us with a more traditional set (with a tuning knob!) they also serve as a chart. As well as your targets, they show the main landmarks and obstacles, and without these you cannot be sure precisely where you are on your small dial.

And like a landscape, it 'looks' different at different times of day or year, and as the weather changes. As evening sets in, we receive some stations much better, others worse. Some close down early. Some are only audible after midnight when stronger stations have gone off the air. Many shortwave stations move to a different number altogether for winter.

Dial-Search can save you hours of frustration and make home listening really rewarding - whether your interest is in music (as a background, or to listen to seriously); or in news from your own area or from 'the old country'; or in a variety of opinions from abroad; or in languages.

And you may find that the 'identification game' becomes a worth-while pastime in itself. If you have never tried it, why not start by switching to 'longwave' and following our one-page list on page 14?

Our medium wave list for Europe is a realistic one that excludes stations we have never heard ourselves with the normal aerial supplied with a set. We indicate the hours each station broadcasts, and the period when (in SE England) it is worth searching for it. Readers in the North, and on the continent of Europe, will also find a brief reference to the other main stations (on the fringes of our map) which they may expect to receive, though our portables here do not.

For this edition I am very pleased to welcome the collaboration of Ian Appleyard, who has revised all the shortwave sections with thoroughness and youthful enthusiasm, including "Broadcasts in English", and has prepared the list of addresses. As usual, all our listings have been up-dated by personal monitoring and research. The book is 8 pages longer; four of these have been needed for the British Isles list, still the most detailed single listing by frequencies on the market, & right up-to-date.

The new British community radio stations have started up at last (p38): a just and welcome development. If local communities can have their own press, their own radio is natural too. For "immigrant languages", too, most European countries already make some kind of radio provision – but how much better if it can be locally based, with residents having access to join in broadcasts! There are also promising newcomers to the London musical scene. The main thing is that all these listed have legality now. May we hope that illegal stations will, in the end, cease to clutter the air? The official French controlling body stated three years ago that "freedom to enjoy clear reception is as important as freedom to broadcast" – a maxim to remember. And I stress again that musicians & their publishers are entitled to royalties on records broadcast.

Of course, if Pirate Gold AM, complete with padre aboard, is your scene, I suppose you are irredeemably anchored on 558. They even contrive to run events on the mainland & advertise them on the air... At least this frequency has regained a certain interest for the dial-searcher late at night, as has 963 all the time. There's an honest new record programme now on 252 which has the courtesy to leave the evening clear. Radio Finland, whose three frequencies these are, must be relieved.

The major Broadcasting Act will soon become law. Though most of the argument has been over its proposals for television, it will affect British radio quite a bit. We the listeners are to have more choice... Medium wave & FM are to provide different programmes (my Dial-Search listing will need a new layout) and the medium wave national frequencies are to be put out to tender soon. All I can advise is to keep an eye & an ear on the news as 1990 progresses, and amend my entries as necessary. One thing is certain: BBC will open its Radio 5 on 27 August, probably on the present MW frequencies of Radio 2; it will carry all the sport & educational broadcasts and a pick of the World Service. There will be an overall Radio Authority, and a Broadcasting Standards Council. The latter idea was unfairly criticised last year as 'puritan, pompous & pointless'. There is such a thing as 'shock radio' (ask the Americans) and some vigilance is required if decadence is to be kept off this precious but most intrusive medium. But not dissent.

Dissent from tradition per se, from the status quo, from the national or local government of the day – over one issue at a time, not from general antagonism – is essential to a thriving democracy. And dissent (<u>or</u> assent) needs to be nourished by information: a fair spread of it, not just what the authorities find it convenient to publish or leak.

The British press & television & radio continue to unearth the information, undeterred by injunctions & other pressure. They are used to getting criticised as too left or too right. (If a news blackout is essential – say to help crime investigation – they will impose one.) BBC radio does a fine job of investigative journalism; if you don't know its regular columns entitled Face the Facts, Punters, File on Four, Analysis, it would be well worth sampling them. In their various ways they have looked into feudal survivals, bureaucratic blockages, conmen, public expenditure – not forgetting the scrupulously fair series My Country Right or Wrong which went on the air, uncut, after a year's interdiction. At this New Year time a prompt from "Twelfth Night" comes to mind: "Your ladyship... you must allow Vox."

Alongside television, BBC Radio 4 serves as a kind of private-eye-cum-ombudsman's-clerk. Parliamentary business too is regularly summarised, including committee work, with taped quotations, and candidly commented on the next morning and at the weekend.

The World Service is equally unblinkered in its programmes about life and institutions in Britain today as well as abroad, and we can learn much from them. Being nominated for an Oscar counts as a distinction, yes? Well, let us record that just two years ago the BBC External Services (as they were called then) were nominated by a group of MPs from all parties for the award of the Nobel Prize for Peace. In 55 years of broadcasting, the citation said, "they have endeavoured to live up to the BBC's motto: Nation shall speak Peace unto Nation. They have contributed to global understanding by ... providing an objective, unbiased, factual and accurate service of world news;" others might well "pursue the same objectives ... in the crucial task of reporting events to mass audiences."

Many listeners find English programmes from abroad refreshing & balancing. 'Europe 1992', a series of short talks from Radio Finland, for instance: last November it told of Scandinavians' amusement at the hesitant progress to Western European Union – they had all had the right, for decades past, to go & work in each other's countries. It is now even more important to learn from the experience/traditions/mistakes of our neighbours: how do they handle such matters as the recycling of waste? house purchase? crime and punishment? transport? financing the arts? manners in parliament? Swiss, French, Swedish, even Scottish practice, can be gleaned from their broadcasts. But material for an overall view is increasingly provided by Radio 4 and the World Service. The 'Europhile' series (R4) was a good addition and ought to become a regular feature.

If you need constant up-dating on the romance of the Dow and April Platinum you will already know when to tune in. Weird in these decimal days to find some bonds still quoted in thirty-seconds (fraction, not time). Mysterious that our loan interest rate should relate to the Deutschmark and that the level of local taxation may depend on the performance of distant municipalities – financial programmes appear to leave all this well alone.

Over recent years the 'phone-in' programme has become widespread in most countries. Often it answers the need for 'company', as does CB radio, or the half outlawed telephone chat-lines. The exchanges may be homely and trivial as on much local radio, or may air personal problems as on some late night programmes, or offer two-way foreign language practice (Hilversum 5), or debate current affairs (Call Nick Ross, BBC R4) or introduce a famous statesman to questioners from all round the globe (BBC World Service). They give listeners far better value than the 'postbag' slots on radio and television, which are far too short & jokily evasive & inferior to the letters columns in the press. Phone-ins also make useful eavesdropping for the foreigner studying our language, or for us studying his.

And surely more Britishers could make the effort to pick up or refresh or up-date their French or German or whatever. Broadcasters often spread quite the wrong attitude by saying a foreign phrase or name and apologising for their dreadful accent, when it is not bad at all! This is not how foreigners come to understand & speak our language: they have a go, they keep at it, they intend to be understood somehow, and they progress. The average Britisher can do the same. The British make excellent linguists or at least resourceful ones, if they want to; it's a question of character. I know English has become a universal language in some spheres (aviation, shipping), but in trade & industry the game will often go to the manager or rep or technician or secretary who can really communicate with his/her European contacts. 1992 is not two years off – in terms of preparation, it was yesterday.

So – listen to European stations; get a brief cassette starter course (plenty at the BBC World Shop or at main booksellers); follow the beginners' courses on foreign radio. Ian's brother learned Swedish from Radio Sweden and now lives & works there. I picked up a basis of three languages from the first courses the BBC put out – besides once learning Urdu in three months because I needed to. Dial-Search originated as a simple guide for students. – Above all, encourage youngsters to see languages as a natural and fun thing to get into. Just the thing for that personal stereo...

1989: what a year of change in Eastern Europe. Radio naturally played a part, though television even more so. Most jamming had already been stopped. Now the broadcasters are noticeably under less strain and constraint. Romania has already changed its signature tune. Content is changing, but programme times are unlikely to.

May I just put on record that in 1988 four people wrote in from the USSR and East Berlin asking for my 5th Edition. They weren't able to send any money, but I sent them the books. I hope they got them. Humbly, I'm glad if Dial-Search has done its tiny bit towards glasnost. Here, too, perhaps?

G.W.
January 1990

4

MAKING THE MOST OF YOUR PORTABLE

Most portable radios are astonishingly selective, simply because it is easy to set them at various angles. A heavier set, or a tuner built into a music centre, does not have this advantage; nor does a set with a fixed external aerial. - If a fixed site has to be chosen for the set, east or west facing is probably best value, in the U.K.

Of course, many stations can be received over a wide angle, & a strong local station at almost any angle. But a portable that remains unturned is wasting at least half of its potential, certainly on the medium waveband.

At least find out where the north is in your home, so that you can turn the set to face approximately the direction indicated by our maps. But it is worth being more precise:-

Take a piece of plain card or stiff paper (A4 or larger) & halve its length by drawing a centre line. Place it with its long side along the front edge of a table or desk which is in a permanent position, say against a wall or in front of a window. Place a compass over the centre line and mark true north and south (accuracy pays; strictly, you should allow for the current magnetic deviation at your location). Draw in the north-south line, and label it "0" at each end. Set a protractor on it and rule "spokes" as in fig.1 on p8 (explained in detail in 5 on p54).

Line the card up again with the table edge, and you have a compass base for your radio set.

The most practical radio is one with a turn-table fitted in the base (such as are still made by Roberts Radio, of West Molesey, Surrey). Stick a paper pointer exactly at the centre of the front, near the bottom. Measure the base of the set and draw its shape on the card, with its centre on the centre of the compass pattern. Get used to placing the set quickly in position on this shape, so that it will rotate uniformly.

The pointer will now give the bearing of reception.

Just a few models of set have the internal ferrite rod antenna placed across the side of the set instead of along the front. In this case follow the advice below but allow 90°.

Controlling our bearing gives us three advantages:-

1. When the set faces the direction given by the map, it will be receiving the maximum signal from the transmitter, giving the loudest and clearest reception. (In some locations, reflection from large buildings &c may distort the angle, in practice.)

2. An unwanted transmitter may be cancelled out ("nulled") by making the end of the set point in its direction, so that reception of this interference (QRM) is at minimum or even zero. Probably you will still be able to receive the other station, the one that interests you. This method may also deal with some electrical interference (QRN).

3. You can often identify an unknown station by nulling it out in this way and noting the bearing of the end of the set in that position. From the map it will be clear whether it is, say, Madrid or Vienna or Paris (all on 585 kHz, 514 m) from this bearing. Turn the set back 90°, for best reception, and listen for announcements, musical clues etc (probably just before the hour) to confirm your identification.

Tuning in on the medium wave band is reasonably easy with a little care and patience, if we choose a station that is not too low-powered or too distant. The "commentary" that follows shows the typical points to listen for. One soon gets experience which can help in more difficult cases of "dial-searching".

The evening brings most variety of stations (and overcrowding too). Suppose we decide to try Swiss Radio for their night programme, giving good concert music from 2300 hours on. The transmitter is at Sottens, some 700 kilometres away from SE England, and is fairly powerful: 500 kilowatts. The frequency is 765 kiloHertz (or 392 metres): see page 17.

We turn the tuning knob until the pointer is between 70 and 80 on the dial. (If the set reads in metres, get it to just under 400). These figures are the "milestones" printed bold in the Dial-Search list.

But there is no way of setting up 765 (or 392) accurately in advance (unless our set has a digital read-out display). There may be only a centimetre of difference between the milestones on the dial. The pointer is a little way below the dial surface and will not look quite the same from all angles (1). The printed dial, or unseen components, may be slightly out (2), or the battery not new. We must rely on our ears now.

Funny, they do have three or four languages in Switzerland, but this sound like Spanish. From page 17 we can tell we are just to one side of the Sottens position. Gently turning the knob (the right way of the numbers!) should correct this. Good, we have German now.

But it's a different sort of programme. If it's near the hour, we could wait for identification: the chirpy four notes of tune 30 on page 48 and "Hier ist der Deutschlandfunk". So back we go, even more gently. With luck, the Swiss programme will come through now quite well, between Germany and Spain. The separation is only 9 kiloHertz each side, on our dial probably less than a millimetre. Before 2300 hours, any speech will be French.

If Sottens is inaudible, check that we haven't got the end of the set pointing at it; there will be minimum reception or none, unless we turn the set (3). Here in Sussex a portable should face more or less south-east (bearing 135°).

Once a month, Sottens is off the air at night for maintenance, and we shall hear the edge of the German programme and the edge of the Spanish one (4). Try next night!

If still silent, let us wait. We may be in a "trough" of fading; the radiated signal may be coming to us by two routes, one fractionally longer than the other, as it is reflected by different layers in the sky (5). The two signals are "out of phase" at one moment but will soon "come together" to give a stronger signal. If this alternation is not too rapid, reception may be acceptable; if not, one can try again later: conditions for "sky-wave" reception vary from hour to hour (and from day to day, with the barometer and with sunspot activity).

Many nights, though, Sottens will come in clear, steady and enjoyable.

At some locations, one also hears a quiet undercurrent of French being spoken, interspersed with popular music. It is not listed for this frequency; it has no business to be there. We worried endlessly over this, but have learned to live with it. The interloper is no pirate; it is Europe Un from the longwave, and its strong signal can combine with others in the same corridor to our receiver (6). Sometimes it is Radio Luxembourg (longwave); sometimes both. The sketch-map, Fig.2, shows the relative positions and the bearings mentioned.

Other interlopers can occur on some rather basic portables, at predictable points (7), and on shortwave even on good sets (8).

October 1986 saw the latest change in the 'landscape' of 765 kHz, when a new British local, BBC Essex, took it for its Chelmsford transmitter. Relaying Radio Two at night, it is clearly heard here (9); though its power is only 500 watts, it is only 110 kilometres away. Luckily it can usually be eliminated by turning the set a little, so that the end of it is pointing at Chelmsford, bearing 8° (just east of north). This does reduce the reception bearing of Sottens from 135° to 98°, but its volume is scarcely affected. A bonus is that one can tune in to BBC Essex with ease, day or night, and the set is then ready tuned to receive Sottens when desired. — On one of my sets, if I then switch to longwave without de-tuning, it happens that I am on 198 kHz; so around midnight this gives me Radio 4 followed by World Service, in addition to the choice of Radio 2 or Switzerland on medium wave by turning the set...

For quick return to a favourite frequency, use a memory button if your set has them. Or if your dial has a 'logging scale' from 0 to 100, read this (1) and record the reading in the margin of Dial-Search. Record the best bearing too, in the column provided. If your set has neither of these facilities, the dial should be marked in some way convenient to you: a wax pencil? a pen? A narrow strip of self-adhesive paper such as typists use for correcting will take many such marks, and can be altered again with correcting fluid.

Almost all stations broadcast on exactly the frequency in kiloHertz that we list. The wavelength in metres is given to the nearest whole number.

Crackles, hums and other noises (10) are likely to be near at hand: switches, motors, TV, fluorescent lights, car & motorcycle ignition etc. They are supposed to have the right 'suppressors' in them to avoid interfering with radio, but these may not be efficient. Luckily most trouble is intermittent or short-lived. If you can work out which is the real culprit in your home, an electrician can put it right. If it is next door or overhead, try different positions for your set. Adding a high aerial will avoid the trouble but also cancel the directional benefits of your portable on medium wave. But a loop aerial will increase these and help against electrical interference too. Thunderstorms, though, will always produce noise (11), even many miles distant.

The footnotes that follow give the official terms for things we have described informally; they are used in more detailed books and in most magazines.

(1) Parallax. (2) Calibration. (3) Orientation.
(4) Adjacent channel interference, or splash.
(5) The ionosphere (some 50 to 400 miles up) which can reflect radio signals over thousands of miles. Shortwave propagation depends on this entirely. Medium and long waves are reflected by it at night; in daytime by a nearer layer, but mainly they are propagated direct by "groundwave".
(6) Ionospheric cross-modulation: the "Luxembourg effect". Another constant example, here in Sussex, is France-Inter from Allouis, heard behind Toulouse on 945 kHz and Tipaza on 254. Readers elsewhere have heard BBC World Service and West German stations on strange frequencies.
(7) Images: transmitter frequency plus 910 kHz (twice the I.F. of usual sets).
(8) Harmonics: usually multiples of the transmitter frequency.
(7 and 8 are "spurious" signals.)
(9) Co-channel interference (QRM).
(10) Electrical interference (QRN). (11) "Static" from lightning, also from space.

The directional effects can be startlingly increased by adding a loop aerial, but this must be at right angles to the set and both must turn together. An old valve radio will have only a loop aerial, wound inside the case; probably the side of the set will need to face the desired station. — Whatever your installation, reception is better the higher your set (or aerial) is. A portable in an upstairs room, near a window if possible, is more reliable than on the ground floor.

VHF transmitters are designed for short range broadcasting, say about 40 km, and many of them are very low-powered. However, from those of 1 kw or more, one may expect moderate reception regularly up to 150 km away, depending on intervening terrain; across the sea, even further. Continental stations can be very clearly received at times, coast to coast, if you are patient; and in exceptional barometric conditions quite long-distance reception is frequently recorded in magazines.

Have your VHF aerial rod fully extended; keep your AFC switched off, also any tone control or muting switch; adjust the tuning very slowly, and try various angles of aerial and set. When searching, the most useful angle for the aerial rod is up at 45°. For any station you stop at, see if reception improves by having the aerial vertical, or (as for many French stations) horizontal. In some cases, less than the full length may be better. Make a note of the best position, also the best bearing. While someone is close to the set, reception may be worsened by their own 'capacitance' (or yours).

Once you have a station clearly tuned in, switch on the AFC. If there is no loss of volume, keep AFC on. If volume is less, switch it off, adjust tuning slightly, and try AFC again. You are correctly tuned when the AFC switch makes no difference to reception; but with it on, it should stay well tuned. If a U.K. national or regional programme seems hard to get clearly, remember it is carried on other frequencies in the same part ot the waveband; you should try to find the strongest and steadiest one, and mark it for future use. Before you start moving on to a fresh station, switch AFC off. — If you haven't "AFC" among your controls, don't worry: your set is designed to cope with it for you.

A "tone" control on a radio, by the way, usually cuts down the treble sounds, reducing hiss if any. By narrowing the bandwidth it may make a difficult station easier to hear. It is not there to make music "sound better", really.

On most sets, VHF needs more volume and batteries are used up rather more quickly.

Shortwave enables you to listen in to broadcasts from every part of the globe. Virtually every country has its 'external services' & these usually include broadcasts in English to Europe, offering detailed news from that country together with comment on international affairs. Many stations aim to give us an idea of what life is like in their country, & offer interesting features on their culture.

The transmissions are beamed out into space but the ionosphere reflects them down to earth, which bounces them up again, & by a series of these bounces the signal can be made to travel vast distances around the curvature of the earth until it hits the 'target area'. The angles will be right for a few hours until the ionosphere alters & a different frequency will be needed: frequencies & schedules are decided by trial & error and long experience. There being only 5 kHz separation between stations on shortwave, the listener needs to tune rather carefully.

On a receiver with digital frequency display (such as the Sony ICF 2001 or its descendants), quickly try the frequencies in a country column in our SW list; a few of these will be carrying a programme from that country, though not necessarily in the language you want. A few others may be in use at that hour by another country — or the adjacent channels may be, and audible! Remember too those MW frequencies that also carry the foreign service in question, useful for comparison.

On more conventional sets, use our SW list horizontally to explore one of the metre-bands. Stop when you hear English spoken (or the language you want); then listen, identify, read the logging scale if any, make notes (including the date and time). Get used to the voices and interval signals from the country concerned.

Like most other guides, we give a timed list of the main broadcasts in English. For other languages, write for schedules to the broadcaster concerned; addresses are given on the air, and in larger books (some on our page 9).

Just use the telescopic whip aerial fitted to your set; for the SW beginner, directional listening is unimportant. When you feel ready to add a "better" antenna advice can be found in leaflets and books (see below) or you could consult your local dealer or radio club.

When the time comes to think of getting a new receiver, think three times before ditching the old one:-

- A domestic portable is designed for quality of sound, and this needs a wide "bandwidth"; it is directional; and it should have VHF(FM), preferably stereo, for best reception of music from your local transmitters;
- A "communications receiver" is designed to be selective at long distances, especially on shortwave, and this requires a narrow bandwidth; it will be adequate for speech but less so for music;
- Ideally, you need one of each, and you may think your old domestic set is still good for its purpose. It can also help with identification, e.g. by having it ready tuned for BBC Radio 2, or France-Inter, or Spanish RNE1, for comparison with other stations you discover on another set.

To keep up-to-date with programmes and times (esp SW), or to understand radio and receivers more thoroughly, the listener needs further reading.

A good Booklist is published (free) by Media Network, Radio Nederland, PO Box 222, 1200JG Hilversum, Holland; also other useful leaflets, inc one on antennas. Listeners can link to their computer, too, for the price of a phone call.

The booklist of the BBC World Shop is also worth sending for (Bush House, Strand, London WC2B 4PH). Interbooks, Perth, Scotland, is a notable specialist bookseller. Your local reference library will have useful titles (not always the latest edition).

BBC & IBA both publish free booklets etc listing all their transmitters and relays and their data. Write to the Engineering Information Dept., BBC, Broadcasting House, London W1A 1AA; or IBA, Crawley Court, Winchester, Hants SO21 2QA.

Radio Times has long had the monopoly to publish full details of BBC programmes; it is expected to lose this soon, so it may become available in other publications. At present the British 'quality' daily newspapers carry quite good coverage, perhaps the fullest and most legible being in The Times. — A monthly summary of the World Service appears in the issue of Radio Times for the first day of the month. Full particulars of W/S programmes & frequencies are in the monthly London Calling (from Bush House; subscription).

Of the many TV magazines in France, Télérama is the most useful for radio, and in W Germany, Hörzu. They give some guide to programmes in neighbouring countries too.

All broadcasting organisations have free leaflets available, detailing their short-wave schedules especially. It is harder to get hold of home service particulars, and they may be wildly out of date!

Monthly magazines about radio are too technical for the layman. Despite its name, Short Wave Magazine now has the most helpful features on broadcasting (all bands). It has a good Book Service too.

Two cassettes are useful: Identification Signals, and Foreign Language Recognition Course, both produced by the Handicapped Aid Programme UK, PO Box 4, St Ives, Huntingdon, Cambridgeshire PE17 4FE.

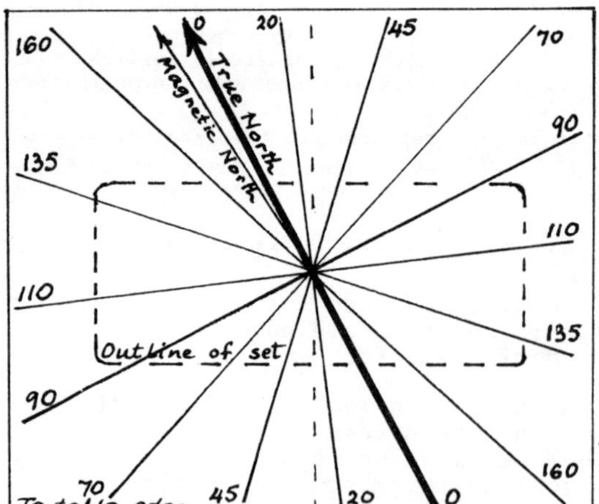

Fig. 1

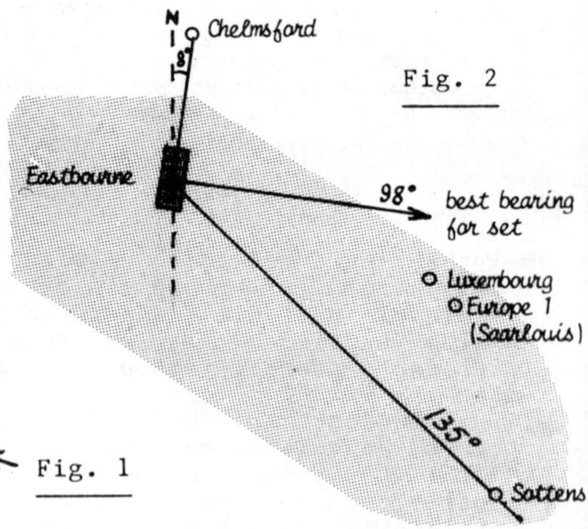

Fig. 2

TIMES of broadcasting in this book are given as British time.

Listeners in most European countries will be one hour ahead of this for most of the year. However, they 'put the clock back' before we do, so that times are the same as the U.K. for a month (in 1990, from Sept. 23 to Oct. 27). This is an interesting period, especially in late evening, as times of close down and fading and interference are different from usual, and clearer reception of some stations can be expected. So:-

> Normally, listeners in Europe should add one* hour to all times printed here (not in October period);
> In October period only, listeners in U.K. and Ireland should add one hour to times printed for European stations and for 'BBC 648';
> Except for the page 'Broadcasts in English'; shortwave broadcasts are always timed in UTC (GMT, i.e. winter time in British Isles) and this also applies to many international service broadcasts on medium wave.

* Local time in Finland & Turkey is 2 hours ahead, in West Russia 3.

It is reported that Britain may line its clocks up with the Europeans, but nothing is decided yet. If it should happen in 1991, disregard all mentions of the 'October period' in this edition.

SOME POPULAR STATION ADDRESSES

Radio Australia, GPO Box 428G, Melbourne, Vic 3001, AUSTRALIA.
Radio Austria International, A-1136, Vienna, AUSTRIA.
Radio Canada International, PO Box 6000, Montreal, CANADA, H3C 3A8.
Voice of Free China, PO Box 24-38, Taipei, Taiwan, Republic of CHINA
HCJB, Casilla 691, Quito, ECUADOR.
Radio Finland, Box 10, 00241, Helsinki, FINLAND.
Radio Budapest, PO Box 1, H-1800, Budapest, HUNGARY.
KOL Israel (Israel Radio Int.) External Services, PO Box 1082, Jerusalem 91010, ISRAEL.
Radio Netherlands International, PO Box 222, 1200JG Hilversum, NETHERLANDS.
Radio Polonia, Al. Niepodleglosci 77/85, 00-950 Warsaw, POLAND.
Radio R.S.A., PO Box 4559, Johannesburg 2000, SOUTH AFRICA.
Spanish Foreign Radio, PO Box 156.202, 28080 Madrid, SPAIN.
Radio Sweden, S-105 10 Stockholm, SWEDEN.
Swiss Radio International, CH-3000, Berne 15, SWITZERLAND.
United Arab Emirates Radio & TV, External Service, PO Box 1695, Dubai, U.A.E.
BBC World Service, Bush House, PO Box 76, Strand, London WC2B 4PH, U.K.
Voice of America, Washington, DC 20547, U.S.A.
Radio Moscow, Moscow, U.S.S.R.

FREQUENCY BANDS

LONG WAVE Low Frequency
LW (*kilometric waves*) l.f.

wavelengths in metres:	frequencies in kiloHertz:
2000 to 1060	150 to 280

MEDIUM WAVE Medium Frequency
MW (*hectometric waves*) m.f.

571 to 187	525 to 1605

SHORT WAVE High Frequency
SW (*decametric waves*) h.f.

including the		kHz
120 metre band (2 MHz band)	(tropical bands)	2300 to 2498
90 metre band (3 MHz band)	(tropical bands)	3200 to 3400
75 metre band (4 MHz band)	(tropical bands)	3900 to 4000
60 metre band (5 MHz band)	(tropical bands)	4750 to 5060
49 metre band (6 MHz band)		5950 to 6200
41 metre band (7 MHz band)		7100 to 7300
31 metre band (9 MHz band)		9500 to 9775
25 metre band (11 MHz band)		11700 to 11975
19 metre band (15 MHz band)		15100 to 15450
16 metre band (17 MHz band)		17700 to 17900
13 metre band (21 MHz band)		21450 to 21750
11 metre band (25 MHz band)		25600 to 26100

Actually many stations now broadcast beyond these limits ('out of band').

VERY HIGH FREQUENCY

VHF	MHz
Band 1: TV (Europe)	41 - 68 MHz
Band 2: VHF (FM) Radio	87.5 - 108
Band 3: TV (Europe)	174 - 230

ULTRA HIGH FREQUENCY

UHF

Band 4:	TV (UK & Europe)	470 - 582
Band 5:	TV (UK & Europe)	614 - 854

SUPER HIGH FREQUENCY

SHF

	GHz
Band 6: TV: DBS (Broadcasting by satellite)	11.7 - 12.5

Barely 10% of the airwaves are allotted to broadcasting (see left). The gaps are used by 'utilities' great & small – land, sea, air & space communications & navigation, emergency & armed services etc. The high-speed noises would need decoding equipment, & listening to the voices is against the law.... but the conversations of transmitting amateurs (hams) may be followed in the upper frequency part of the AMATEUR BANDS detailed below:-

160 m band	1850 - 2000 kHz
80 m band	3500 - 3800
(75 m band	Americas only)
40 m b	7000 - 7100
(41 m b	Americas only)
30 m b	10100 - 10150 [a]
20 m b	14000 - 14350
17 m b	18068 - 18168
15 m b	21000 - 21450
13 m b	24870 - 24990
27 MHz band: CB [b]	
10 m b	28 MHz - 29.7 MHz

[a] Morse (CW) & teletype (RTTY) only.
[b] Citizens' Band. FM only, in UK.

(Normal receivers stop here, adding VHF Band 2)

4 m b	70 MHz
2 m b	144 MHz

934 MHz band CB(FM). UK only.

& bands at 2.3, 3.4, 5, 10 & 24 GHz

The frequency or wavelength is that of the transmitter's steady 'carrier wave'; the sound or vision is 'modulated' on to this by amplitude modulation (AM), except that in VHF radio it is done by frequency modulation (FM), which allows stereo transmission. Standard frequency units are:-

1 kiloHertz (kHz) = 1000 cycles per second (1 kc/s in the old units). To save space on dials, zeros are often left off, e.g. 530 kHz shown as 53 or even 5.3. 1000 kHz = 1 MegaHertz (MHz). Shortwave frequencies are often given in MegaHertz, e.g. 11700 kHz = 11.7 MHz.
And 1000 MHz = 1 GigaHertz (GHz).

Divide 300 000 by the frequency in kHz, to obtain the wavelength in metres, if needed; divide 300 000 by wavelength in metres, to obtain frequency in kHz.
Easy, on a calculator!

ABBREVIATIONS

24h = continuously on the air.
06 = 0600; 13 = 1300 hrs (1 p.m.)
24 = 2400 hrs, 00 = 0000 hrs (midnight.
GMT = Greenwich Mean Time.
UTC = Universal Time Coordinated.
M = Monday. T = Tuesday.
W = Wednesday. Ŧ = Thursday.
F = Friday. S = Saturday.
$ or s = Sunday.
R = Radio, or Russian.
G = German. Fr = French.
Sp = Spanish. W/S = World Service.
c = approximately.
esp = especially.
usu = usually.
occ = occasionally.
inc = including.
* = planned.
Prog or Pr = programme.
RT = Radio (&) Television.
RF = Rundfunk (= broadcasting).
AFN = American Forces Network.
AFRTS = American Forces R & TV Service.
AFC = automatic frequency control.
RIAS = R im Amerikanischer Sektor.
DDR = Deutsche Demokratische Republik.
BDR = Bundesrepublik Deutschland (West).
DLF = Deutschlandfunk.
RNE = R Nacional de España.
Other abbreviations are explained where first mentioned.

Medium Wave from North America

Some listeners find this an intriguing challenge. It does involve staying up late, for there is no chance of reception in Europe until we have had five hours of darkness here. By then the eastern seaboard of N America is dark too, & care & patience & a good aerial may be rewarded. Frequencies are 10kHz apart (not 9 as in Europe) starting at 540 kHz & ending at 1600. Stations identify by callsign; programmes are in English. This is just a selection of the ones most frequently logged:-

590 kHz (508 m) VOCM St John's, Newfoundland (10 kw)

600 (500) CBNA St Anthony, Nfd (10)

610 (492) CKYQ Grand Bank, Nfd (10)

710 (423) CKVO Clarenville, Nfd (10)

770 (360) WABC New York (50)

880 (341) WCBS New York (50)

930 (325) CJYQ St John's, Nfd (25)

1010 (297) WINS New York (50)

1030 (291) WBZ Boston (50)

1050 (286) WFAN New York (50)

1070 (275) WBAL Baltimore (50)

1110 (270) CBD St John, New Brunswick (10)

1130 (266) WNEW New York (50)

1210 (250) WCAU Philadelphia (50)

1500 (200) WTOP Washington DC (50)

1510 (199) WSSH Boston (50)

1560 (192) WQXR New York (50)

1570 (191) CKLM Laval, Quebec (50) Fr

And from further south:-

570 (526) R Rumbos, Caracas (100) Sp
1100 (273) R Globo, Sao Paulo (50)
1220 (246) R Globo, Rio de J. (150)
(both Portuguese)

1610 (186) Caribbean Beacon (evangelical), Anguilla (50)

INDEX of Programme Networks

Covering all MW & LW stations listed in bolder type on pp 14 - 23, grouped according to main language.
! = other languages are included. — Refer to p43 and main lists for fuller details.

ENGLISH	kHz	m
R4	LW 198	LW 1515
R3 inc sport	1215	247
R2 inc sport	693, 909	433, 330
R1	1053, 1089	285, 275
R Scotland	810	370
R Wales	882	340
R Ulster	1341	224
World Service	648! LW 198 1296!	463! LW 1515 231!
RT Eireann 1	567, 729	529, 412
RTE 2	612	490
Atlantic 252	LW 252	LW 1190
RT Luxembourg	1440!	208!
Amer Forces Netwk	873, 1107, 1143	343, 271, 263
Voice of America	1197!	251!
British locals <E>	see pp 24 to 38	
DUTCH Hilversum 3	675	444
Hilversum 1/2	747	401
Hilversum 5	1008	298
Belgium: BRT 1	927	324
BRT 2	540, 1188	556, 253
+ International	1512!	198!
DANISH Prog 1	LW 243!	LW 1224!
Prog 3	1062	282
SWEDISH Pr 1	1179! LW 189	254! LW 1587
NORWEGIAN Pr 1	630, 1314,	476, 228,

FRENCH	kHz	m
France Inter	LW 162; 675, 1071, 1161, 1350	LW 1852; 444, 280, 258, 222
R France MW (!) (p43)	711, 837, 864, 945, 1206, 1278, 1377, 1404	422, 358, 347, 317, 249, 235, 218, 214
FIP	585	514
Europe 1	LW 183	LW 1639
R Monte Carlo	LW 216	LW 1388
RMC + TWR	1467!	204!
RT Luxembourg	LW 234	LW 1282
Sud Radio	819	366
Algeria	981! LW 252	306! LW 1190
Tunisia	963!	311!
R Suisse Romande	765!	392!
RTBF 1	621	483
RTBF 2	1125	267
RUSSIAN: R Moscow	LW 171; 1143, 1323! 1386! 1494!	LW 1754; 263, 227! 216! 201!
R Volga	LW 261	LW 1149 /317
Mayak Pr:	LW 198, 261, 279; 945	LW 1515, 1149, 1075;/
Ukraine	1377, 1404	218, 214
Vilnius	1557	193
POLISH Warsaw Pr 1	LW 225	LW 1333
Prog 2:	819, 1080, 1206, 1305	366, 278, 249, 230
+ R Polonia	1503!	200!
BULGARIAN		
Sofia 1, Horizont	963, 1224!	311, 245!

FINNISH	(!)	558, 963, LW 252	538, 311, LW 1190
ITALIAN	R 1	657,819,900, 1332, 1575	456,366,333, 225, 190
	R 2	846,1449	355, 207
Vatican R		1530!	196!
Swiss		558	538
R Monte Carlo		702	427
SPANISH			
RNE 1:		558,585,621,639, 684,729,738,774,855	538,514,483,468, 439,412,406,388,351
Others:		657,792,828,837, 918,990,999,1260,1413	456,379,362,358, 327,303,300,238,212
PORTUGUESE	Pr 1	666 450	450
	Pr 3	1035	290
Others		594,720,963	505,417,311
ROMANIAN	Bucharest 1:	LW 153	LW 1960
Bucharest 2		756!	397!
HUNGARIAN			
Prog 1, Kossuth R		540	556
Prog 2, Petöfi R		1188,1251	253, 240
ALBANIAN (!)	Tirana	1206, 1215,1395,1458	249, 247,215,206
ARABIC	Egypt	621	483
Algeria		531,549,891	565,546,337
Morocco		540,612!711,LW 207	556,490!422;1449
(Berber pr)		702	427
Tunisia		630	476
Libya		1251	240

CZECH/SLOVAK*	Praha	639	468
Hvezda		LW 270; 1233,1287!1521	LW 1111; 243,233!197
* Bratislava		702,1098	427,273
YUGOSLAVIA			
Belgrade 1		684	439
Zagreb 1		1125 to 1143	267 to 263
Ljubljana 1		918	327
Sarajevo 1		612	490
Pristina 1		1413	212
GERMAN	DLF	LW 153,207; 549,756, 1269!1539!	LW 1960,1449; 546,397, 236!195!
Deutsche Welle		1557!	193!
Bavaria		801	374
Bremen		936	320
Hesse		594	505
NDR		630,972	476,309
Saar		1422	211
Südfunk (SDR)		576,711	520,422
SWF		666,1017	540,295
WDR		1593	188
RIAS		855,990	351,303
DDR 1		783,1044	383,287
Voice of DDR		LW 177	LW 1695
" " "	+ RBI	1359!	221!
Austria	ORF 1	585,1026	514,292
	ORF 1 + 3	1476	203
Swiss	DRS 1	531,1566	565, 191
RT Luxembourg		1440!	208!

LONGWAVE

The data listed are (from left to right):- frequency in kiloHertz; wavelength in metres; a space for your bearing (see p54 & p4); languages used (capitals indicate strongest reception in SE England, brackets the weakest or least reliable); name of programme; name of station or transmitter and its lowest power in kilowatts; broadcasting hours (underlining shows when usually audible here). All timings are British time (see p9). At right, other major stations on same frequency, but not logged by us, or seldom.

kHz	m	bearing	Station	Others
153	1960		(German): DLF (Donebach 250) 24h	+ R 500
			(Romanian): Bucharest 1 (Brasov 1200) 24h	+ Algeria 1000
162	1852		FRENCH: France Inter (Allouis 2000) 24h: closed T 00-04	+ Turkey 1000
171	1754		(Russian): R Moscow Prog 1 (Kaliningrad 1000 &c) 02-08-15-22-01	+ Morocco 2000
177	1695		(German): Voice of the DDR (Oranienburg 750) 24h	
183	1639		FRENCH: Europe 1 (Saarlouis 2000) 05-02, S-§ 24h	+ 180 Turkey 1200
189	1587		(Swedish): Prog 1 (Motala 300) c 0430-18-21- c 2230	+ R 500
198	1515		ENGLISH: BBC Radio 4 (Droitwich 500 &c) 0550-0045	+ Algeria 1000
			" BBC World Service 0045-0550	+ Poland 200
			(Russian): Mayak Prog (Leningrad 150 &c) 03-02	
207	1449		(German): DLF (Munich 500) 24h 17-01	
			(Arabic): Morocco (Azilal 800) 24h 02-04	
216	1388		French: R Monte Carlo, Monaco (Roumoules 1400) 04-2305	
			(Norwegian): Prog 1 (Oslo 200) 24h 16-04	
225	1333		Polish: R Warsaw Prog 1 (Konstantinow 2000) 24h	
234	1282		FRENCH: RT Luxembourg (Junglinster 2000) 0430 (§ 05) - 04	+ R 1000
243	1234		Danish: Prog 1 (Kalundborg 300) 05-2220.	
252	1190		French: Algeria, Chaîne 3 (Tipaza 1500/750) 05-06-19-0005	
			Finnish &c: Finland (Lahti 200) 24h 0005-05 (Engl 0005, Fr 0040)	
			Music & English: Atlantic 252 (RTE & RTL) (Summerhill 0.1/0.5) 06-19	

261	1149	Russian: R Volga (Burg 200) 05-09-22 then Mayak Pr (Moscow 2000)
270	1111	Czech: Hvezda Prog (Topolná 1500) 24h
279	1075	(Russian): R Moscow Pr 1 + Mayak Pr (Minsk 500) 02-22

MEDIUM WAVE

Blank lines marked <E> are for your notes of U.K. stations heard (full list pp 25-38); only the most powerful are printed in the list below.

kHz	m	bearing		
531 53	565		Arabic: Algeria 1 (Ain Beida 600/300) 24h 17-02	
			(German): Swiss, DRS 1 Beromünster 500) daytime only	+ DDR 100
540	556		DUTCH: Belgium, BRT2 (Wavre 150/50) 0430-01	
			Arabic: Morocco (Sidi Bennour 600) 05-17-01	+ Kuwait 1500
	550		(Hungarian): Pr 1, Kossuth R (Solt 2000) 0330-19-0320	
549	546		German: DLF (Bayreuth 200, Nordkirchen 100) 24h	+ R 1000
			(Arabic): Algeria 1 (Oran, Les Trembles 600/300) 24h 18-02	+ Saudi 2000
558	538		<E>	
			Italian: Swiss, Rete 1+3 (Monte Ceneri 300) 24h 05-06-22-05	
			(Spanish): RNE1 (?) 24h 19-04	
			(Finnish &c): Pr 1 (Helsinki 100) 24h 22...	
567	529		ENGLISH: RTE R1 (Tullamore 500) 0630-24	+ W Berlin 100
576	520		(German): Sudfunk 1 (Mühlacker 500) 0425-09-2305	+ DDR 250 + R + Algeria
585	514		German: Austria, ORF1 (Vienna 600/240) 05-16-2305 (S 24) inc Ö Reg S 19-21	
			Spanish: RNE1 (Madrid 200) 24h 17-04	
			Music & Fr: FIP (Paris 8) 07-15 or 17	+ Tunisia 350 + Saudi 1200
			<E>	
594	505		(German): Hessischer RF (Frankfurt-am-Main 400 &c) 24h 15-03	+ Bulg 250
			(Portuguese): RRE (Muge 100) 24h 20-02	+ Saudi 2000

kHz	m	Station	Also
603 (60)	498 (500)	<E>	+ Radio France Lyon 300 + DDR 30
612	490	English: RTE R2 (Athlone 100) 0630-12-17-0150	
		Arabic (& Fr): Morocco, R Medi 1 (Sebaa Aioun 300) 06-20-24	+ Spain
		(Serbo-Croat): Yugoslavia, Sarajevo 1 (600) 04-18-21-24	+ R (Tallinn)
621	483	French: Belgium, RTBF1 (Wavre 80) 0430-2245	
		(Arabic): Egypt, various prs (Batra 2000) 03-22-01	
		(Spanish): R España (Barcelona 10) or RNE (Tenerife, Canary Is 50) 24h 20-02	
630	476	(Arabic): Tunis (Djedeida 600) 0430-17-2330	+ Turkey 300
		(Norwegian): Prog 1 (Vigra 100) 24h 23-03	+ Romania 400
		(German): SFB/NDR (Dannenberg 80) c 05-23	
		<E>	
639	468	Spanish: RNE1 (La Coruña 100 &c) 24h 16-02	+ Cyprus
		(Czech): Praha Prog (Liblice 1500) 0330-10-23	(BBC W/S 500)
648	463	ENGLISH (+ Fr & G, see p26): BBC World Service/BBC 648 (Orfordness 500) 24h	+ R 150 + Saudi 2000
657	456	Spanish: RCE, RC2 (Madrid 20) ...17-03...	+ DDR 250
		(Italian): Radio 1 (Naples 120 &c) 05-16-2230	+ R 150
		<E>	
666	450 (450)	(Portuguese): Prog 1 (Lisbon 135) 24h 19-02	+ Spain
		(German): Südwestfunk (Rohrdorf 180) 0430-23	+ Vilnius 500 (R)
		<E>	
675	444	DUTCH: Hilversum 3 (Lopik 120) 06-23	
		(French): France Inter (Marseille 600) 24h 16-03	
684	439	Spanish: RNE1 (Seville 250) 24h 15-02	
		(Serbo-Croat): Yugoslavia, Belgrade 1 (2000) 24h 17-01	
693	433	ENGLISH: BBC Radio 2 (Droitwich 150 &c) 24h	+ Berliner RF 250

kHz	m	Programme	Also
702	427	(Italian): R Monte Carlo (300) 0530-16-1835 + Fr, TWR	
	70	(Berber): Morocco (Sebaa Aioun 240) 06-22-01	+ G (BRD) 5
		(Slovak): Bratislava Pr (Presov, Banska Bystrica 400) 03-16-23	+ Sp
711	422	FRENCH &c (p43) 0530-22 (Rennes 300): Radio Bleue 06 - 18 / Radio France Armorique / RFI	
		(Arabic): Morocco(600) / Egypt (100) ...19-01...	+ R 150
		(German): SDR 1+3 (Heidelberg, Ulm &c 5) 24h 16...24	+ Yugo 20
720	417	(Portuguese): Antena 1 (Norte Azurara 100) 24h 21-03	+ WDRF 200
		<E>	+ RFE Holzkirchen
729	412	English: RTE R1 (Cork 10) 0630-21-0150	
		(Spanish): RNE1 (Oviedo 50 &c) 24h 19-02	+ Greece 150
		<E>	
738	406	Spanish: RNE1 (Barcelona 250) 24h 18-03	+ RFI (Paris 4) + Poland
		<E>	
747	401	DUTCH: Hilversum 2 (Flevoland 400) 06-23	+ Bulgaria 500
	400		
756	397	German: DLF (Brunswick 800/200) 24h 11-03	
		(Romanian: evening, Balkan languages): Bucharest 2 (Lugoj 400) -20-24-	
		<E>	
765	392	French (&c:p43): R Suisse Romande 1 (La Première) (Sottens 500) 24h inc Espace 2 (19-05) 17-07	+ R 150
		<E>	
774	388	Spanish: RNE1 (Cáceres 60, San Sebastian 50 &c) 24h 16-02	+ Egypt 500
		<E>	+ R 100
783	383	(German): DDR1 & Voice of DDR (Burg 1000) 24h 16-04	+ Portugal 50
792	379	Spanish: SER, R Sevilla (20) 24h 16-03	+ R Fr Limoges 300
		<E>	+ Greece 500 + Egypt

kHz	m	Station	Also
801 **80**	374	(German): Bayerischer RF (Munich 420) 0430-15-24	
		<E>	+ R 500
810	370	(English): BBC R Scotland (Westerglen 100 &c) 06-14-0045	+ Sp + R + Yugo
819	366	(Polish): Pr 2(Warsaw 300) 0430-20-23	
		(Italian): R 1 (Trieste 25) 05-18-2230	
		(French): Sud Radio (Toulouse 50 &c) 24h 16-19	+ Sp
		<E>	
828	362	(Spanish): SER, R Barcelona (20) 24h 19-03	+ NDRF 100
		<E>	+ Bulg 500 + Libya 300
837	358	(Spanish): COPE, R Popular Sevilla (10) 24h 19-03	+ R 150
		(French &c) p43 0530-18-22 (Nancy 200): R Bleue 06 - 18 / RFI	
		<E>	
846	355	ITALIAN: R 2 (Rome 540) 24h 15-03	
855	351 **350**	Spanish: RNE1 (Murcia 125) 24h 15-02	
		(German): RIAS (W Berlin 100) 0325-19-0135	+ Romania 1500
		<E>	
864	347	French &c (p43) 0530-18-2030-22 (Paris 300) R Bleue 06-18/RFI	+ Egypt 500 + R + Bulg
873	343	(English): American Forces Network (Frankfurt 150) 24h 17-03	+ Sp 20 + R 150
		<E>	
882	340	English: BBC R Wales (Washford 100) 24h	+ DDR 250 + Yugo 300 + R 150
891	337	Arabic: Algeria 1 (Algiers 600/300) 24h 15-03	+ Neths 20
900 **90**	333	Italian: Radio 1 (Milan 600) 24h 17-03	+ Czech 30 + Saudi 1000
909	330	ENGLISH: BBC Radio 2 (Moorside Edge, Brookmans Park 150 &c) 24h	
918	327	Slovenian: Yugoslavia, R Ljubljana 1 (600/100) 24h 17-23	
		(Spanish): R Intercontinental (independent, Madrid 20) 24h 22-04	

kHz	m	Station	Other
927	324	DUTCH: Belgium, BRT1 (Wolvertem 300) 0430-10-2230	
936	320	German: R Bremen 1 (100) 24h 17-02 <E>	+ Italy 10 + Morocco 600 + R 500
945	317	French &c (p43) 0530-16...23 (Toulouse 300): R Bleue 06 - 18 / RFI (Russian): Mayak Pr (Riga 50, Rostov 300) 02-20-01 <E>	
954	314	<E>	+ Sp + Czech 400 + Turkey 300
963	311	Finnish + E, Fr, G, Sw: Pr 1 + 2 + W/S (Turku 600) 24h 14-01: Engl 0730,1930,22,24 (French: Ital 1330-1430: Arabic 22) Tunisia (Tunis 200) 05-21-01 (Portuguese): RRE (Seixal 10) 24h 19-04 (Bulgarian): Pr 1, Horizont (Sofia 150, Shuman 50) 24h 18-01	+ Ireland
972	309	German: Norddeutscher RF (Hamburg 300) 24h 15-03	+ R 500
981	306	French &c: Algeria, Chaîne 3 (Algiers 600/300) 05-17-0005 (English 19-20: Spanish 11-12; 20-21, 02-0230)	
990	303	German: RIAS (W Berlin 300) 24h 15-03 (Spanish): SER, R Bilbao (10) 24h 18-03 <E>	
999 **100**	300 **300**	Spanish: COPE, R Popular Madrid (50) 24h 20-03 <E>	+ R 100 + Malta 20
1008	298	Dutch: Hilversum 5 (Flevoland 400) M-S 08-2130 **(immigrant langs 18-19)**	+ Yugo + Sp
1017	295	German: Südwestfunk (Wolfsheim 600) 24h 18-02	+ Turkey 1200
1026	292	(German): Austria, R Vorarlberg (Dornbirn 50) 05-17, ORF 1 (Linz 100, Dornbirn 50) 17-19-2230 inc S 19-21 Ö Regional <E>	+ Sp
1035	290	(Portuguese): Prog 3 (Lisbon 135) 24h 19-03 <E>	+ Italy R2 50 + R 500

kHz	m	Station	Also
1044	287	(German): DDR1 (Burg &c 250) 24h	+ Morocco 300 + Greece 150
1053	285	ENGLISH: BBC Radio 1 (Droitwich 150 &c) 05-02	+ Romania 1000 + Morocco 600
1062	282	Danish: Pr 3 (Kalundborg 250) 04-19-2305	+ Italy + Portugal + Turkey
1071	280	French: France Inter (Lille 40, Brest, Grenoble, Bastia 20) 0430-21	+ R 60
1080	278	Polish: Prog 2 (Katowice 1500) 0430-16-24	
1089	275	ENGLISH: BBC Radio 1 (Brookmans Park 150 &c) 05-02	+ R 300 + Alb 150
1098	273	Slovak: Bratislava (400) 03-16-02	
1107	271	(English): American Forces Network (Munich 40 &c) 24h 18-02 <E>	+ Egypt 600 + Sp + Yugo 150 + R 150
1116	269	<E>	+ Italy R2 150
1125	267	French: Belgium, RTBF2 (La Louvière 20) 0430-10-2245 Serbo-Croat, Yugoslavia, R Zagreb 1, 04-16-23 (S 24) (Tovarnik 300) 24h	+ R 150 + Bulg 500 + Libya 500
1134	265	Serbo-Croat, Yugoslavia, R Zagreb 1, 04-16-23 (S 24) (Zadar 1200)	+ Sp
1143	263	Serbo-Croat, Yugoslavia, R Zagreb 1, 04-16-23 (S 24) (Zagreb 135) (Russian & many): R Moscow (Kaliningrad 150) 24h 19-03 (English): American Forces Network (Stuttgart 10 &c) 24h 21-24	
1152	261	<E>	+ Romania
1161	258	<E> (French): France Inter (Strasbourg 200, Toulouse 100) 24h	+ Bulgaria
1170	257	<E>	+ R 500 + DDR + Sp
1179	254	SWEDISH & many: R Sweden (Sölvesborg 600) 0430-14-24	+ Romania + Sp
1188	253	(Hungarian): Pr 2, Petöfi R (Szolnok 135) 0325-19-2325 Dutch: Belgium, BRT2 (Kuurne 5) 0430-10-01	+ DDR

kHz	m	Station	
1197 **120**	251 **250**	English & E European languages: Voice of America (Munich 300) 11-14-19-07 (English 11-1630,17-1730,00-02,0430-05,06-07) <E>	
1206	249	(Polish): Pr 2 (Wroclaw 200 &c) 0430-16-24 (Various): Albania (Korce 10) (French &c) p43 0530-21-22 (Bordeaux 100): R Bleue 06 - 18 / RFI	
1215	247	(English): BBC R3 (Moorside Edge 100 &c) 0655-24 (Various): Albania (Lushnje 500) ...20-23 then R Mayak Pr	/ + Sp
1224	245	Bulgarian & (evening) Balkan languages: Sofia 1, Horizont (Vidin 500) 24h 16-22	
1233	243	Czech: Hvezda Pr (Prague 400) 24h 14-03	+ Morocco 200 + Cyprus 600 + RTBF2
1242	242	<E>	+ R-Fr Marseille 150 + Kiev 150
1251	240	Arabic: Libya (Tripoli 500) 05-16-04 (Hungarian): Prog 2, Petöfi R (Marcali 500) 0325-19-2325 <E>	+ Neths
1260	238	Spanish: SER (Valencia 20 &c) 24h 16-24 <E>	+ VoA Greece 500 + Spain 20
1269	236	German (evening, English, Swedish &c): DLF (Neumünster 600) 24h 14-04	+ Yugo 600
1278	235	(French &c) p43 0530-14-19...21-22 (Strasbourg 300): R Bleue / RFI <E>	+ R 150
1287	233	Czech & many: Czech W/S (Litomysl 300, Liblice 200) 24h 15-03	
1296	231	(East European languages + G + few periods English): BBC (Orfordness 500) 0345-08, 09-10, 1215-13, 1615-17, 1930-23	+ Bulgaria 150 + R 150
1305	230	Polish: Prog 2 (Rzeszow 100 &c) 0430-18-23 (French): RTBF 2 (Marche 2) 0430-19-2245 <E>	+ Italy

kHz	m	Programme	
1314	228	Norwegian: Prog 1 (Kvitsøy 1200) 24h 11-23	+ R 250 + Yugo 100 + Sp 10
1323	227	Many: R Moscow W/S (Leipzig 150) 24h 16-03 <E>	+ BBC W/S Cyprus 200
1332	225	Italian: Radio 1 (Rome 300) 05-15-2230 <E>	
1341	224	English: BBC R Ulster (Lisnagarvey 100) 24h 16-24	+ Hungary 300
1350	222	French: France Inter (0430-21 Nancy 100; 24h Nice 300) 09-03	+ R 150
1359	221	(German & many): Voice of DDR + evenings RBI (E Berlin 250/100) 24h 18-03 <E>	+ R 150 + Alb 50
1368	219	<E>	+ Poland 60
1377	218	FRENCH &c (p43) 0530...22 (Lille 300): R Bleue 06 - 18 / RFI (R / Ukrainian): Pr 2 (Ukraine &c 50) 24h 17-03	
1386	216	Many: R Moscow Foreign Service (Kaliningrad 500) 02(R)-15-01	
1395	215	Albanian & many: R Tirana (Lushnje 1000) 04-15-24	
1404	214	(French &c) p43 0530-16-19-22 (Brest, Pau, Ajaccio 20: Dijon, Grenoble 1): R Bleue/RFI (R/ Ukrainian): Pr 2 (Lvov 30 &c) 24h 15-03	
1413	212	(Spanish): RCE (Saragossa 20, Seville 10 &c) ...20-24... (Serbo-Croat): Yugoslavia, R Pristina 1 (1000) 04-23 (24h S §) 21-03	
1422	211	(German): Saar, SRF1 + ARD (Heusweiler 600+) 24h 16-03	
1431	210	<E>	+ R 400
1440	208	GERMAN &c: ENGLISH 1845-03: RT Luxembourg (Marnach 1200) 0430-03 (p31)	
1449	207	(Italian): R2 (Turin &c 20/50) 05-18-2230 <E>	
1458	206	Albanian & many: R Tirana (Lushnje 500) 04-14-24 <E>	

kHz	m	Details
1467	204	FRENCH: evenings MANY, TransWorld R, evangelical: R Monte Carlo (Roumoules 1000) 06-15-0015
1476	203	German: Austria, ORF 1 or 3 (silent 06-18) inc S 19-21 Ö Regional (Vienna 600) + Sp <E>
1485	202	<E>
1494	201	Russian & many: Foreign Service (Leningrad 1000) 1630-19-02/06 + R Fr (Bastia, Clermont)
1503	200 **200**	Polish & many: Warsaw Home & W/S (Stargard 300) 04-16-01 + Sp <E>
1512	198	Dutch (0430-18) Belgium, BRT2 (20kw) + Studio Brussel M-F 15-18; Many (18-2245): BRT International (Wolvertem 300)
1521	197	Czech/Slovak: Hvezda Pr (Kosice 600) 24h 14-01 + Saudi 2000 <E>
1530	196	Italian & many: Vatican City (Rome 150/450) 03-14-24 <E>
1539	195	(German: evenings English, Fr &c): D L F (Mainflingen 700) 24h 16-03
1548	194	<E>
1557	193	<E> (English &c): Malta (Cyclops 600): R Mediterranean, Arabic 18-19, Fr 2130, Engl 2230-2330; & Deutsche Welle, G or Arabic, mainly 12-18 + R Fr Nice 300 (R/Lithuanian): R Vilnius (Kaunas 75) ...21-23 + Yugo 20
1566	191	(German): Swiss, DRS 1 (Sarnen 300) 05-07 & 16-17-21-24 + Tunis 1200
1575	190	(Italian): Radio 1 (Genoa 50) 05-15-2230 + Portugal 10 + Berliner RF 250
1584	189	<E> + Spain
1593	188	GERMAN: Westdeutscher RF Pr 1 (Langenberg 400/800) 24h 16-24 + Portugal 100 + Cz
1602	187	<E>
1611	186	+ Vatican

INDEX of British Local Stations

Figures are in kiloHertz; refer to MW list on pp 25-38 for full details, alternative frequencies, and VHF(FM) data.
B = BBC. I = Independent. Rg = BBC National Regions. R = Radio.

Barnsley I 1305
Barnstaple B 801
Barrow B 837
Bath I 1260
Bedford B 1161, I 792
Bexhill B 1161
Birmingham B 1458, I 1152
Bodmin B 657
Bournemouth B 1359, I 828, R1 1485, R3 1197
Bradford I 1278
Brighton B 1485, I 1323
Bristol B 1548, I 1260
Bury St Edmunds I 1251
Cambridge B 1026, R3 1197
Canterbury B 774, I 603
Carlisle B 756, R4 1485
Chelmsford B 756, I 1359
Chippenham B 1332, I 936
Cleveland B 1548, I 1170
Colchester B 729
Cornwall B 630
Coventry I 1359
Crawley B 1368, I 1521
Cumbria B 756
Derby B 1116, I 945
Devon B 801, I 666
Doncaster I 990
Essex B 729, I 1431
Exeter B 990, I 666
Furness B 837
Gillingham B 1035, I 1242
Gloucester B 603, I 774
Guildford I 1476
Hereford B 819, I 954
Huddersfield I 1530
Hull/Humberside B 1485, I 1161
Ipswich I 1170
Kent B 774, I 603
King's Lynn B 873
Lancashire B 855, I 999
Lancaster B 1557
Leeds B 774, I 828
Leicester B 837, I 1260
Lincoln B 1368
Liverpool/Merseyside B 1485, I 1548, R1 1107
London R4 720, B 1458, I 1152, 1548
Luton B 630, I 828
Manchester B 1458, I 1152
Mansfield B 1584
Newcastle B 1458, I 1152, R4 603
Norfolk/Norwich B 855, I 1152
Northampton B 1107, I 1557
Nottingham B 1521, I 999
Oxford B 1485
Peterborough B 1449, I 1332
Plymouth B 855, I 1152, R4 774
Preston B 855, I 999
Portsmouth I 1170
Reading I 1170
Redruth B 630, R4 756
Reigate I 1521
Rotherham I 1548
Scarborough B 1260
Sheffield B 1035, I 1548
Shrewsbury/Shropshire B 756, I 990
Southampton/Solent B 999, I 1557
Southend B 1530, I 1431
Stockton-on-Tees B 1548, I 1170
Stoke-on-Trent B 1503, I 1170
Sussex B 1485, 1161, I 1323
Swindon B 1368, I 1161
Taunton/Somerset B 1323
Tavistock I 1152
Torbay B 1458, I 954, R3 1197
Tunbridge Wells B 1602
Whitehaven B 1458
Wiltshire B 1368, I 936, 1161
Worcester B 738, I 1530
Wolverhampton B 828, I 990
(Great) Yarmouth I 1152
York B 666
Wales Rg 882, 657, R2 990
Cardiff I 1359
Swansea I 1170
Wrexham Rg 657, I 1260
Scotland Rg 810, 585, 990
Aberdeen I 1035, Rg 990, R4 1449
Ayr I 1035
Dundee I 1161
Edinburgh I 1548
Glasgow I 1152
Inverness I 1152
Perth I 1584
Northern Ireland I 1026, R3 1197, R4 720, 774, Rg 1341, 792, 873
Channel Islands B 1026, 1116
Isle of Man 1368

MEDIUM WAVE (BRITISH ISLES)

For United Kingdom frequencies this is a complete list. The Republic of Ireland is also included (RTE & a few locals) also Luxembourg. Alternative MW freqs are shown here for cross-reference in metres only, these numbers being well-known in titles, jingles, phone numbers &c. STD phone numbers, to "listen to output" of a station, have been discontinued recently. Broadcasting hours are shown up-dated to Jan.1990; liable to alteration. All BBC local stations are 24h; late evening, 6 or 8 stations usu take one regionally produced prog (shown as Reg) before going over to R2 (or R1) for the night.

Ind = commercial stations, still under the IBA, due to be replaced in 1990 by a new Radio Authority.

BBC = British Broadcasting Corporation; a number of its MW freqs are soon likely to go to new independent national & local stations, but bids have not yet been invited so details are impossible to guess.

VHF antenna polarisation: H = horizontal, V = vertical, M = mixed.

At end of this list: Stations on VHF/FM only, and Community Radio stations.

kHz	m			also on (metres)	VHF(FM): MHz	kw	AE	
198	1515	LW	BBC Radio 4 (Droitwich 500, Westerglen 50, Burghead 50)	498,397, 388,417, 202	92.4 to 94.6 (Channel Is 97.1)			0550-0045, then World Service to 0550
252	1190	LW	Atlantic 252 (Summerhill 0.5 0.1)		-	-		06 - 19; popular music. Sponsored RTE & RTL.
531	565	MW	Private: Sunshine R (Dublin 5)		100.5,101.0,101.5			24h
540	556		R na Gaeltachta (Conamara 2)	362,311				See 311m
558	538		Private: R Caroline (off Essex coast: 7.5)		-	-		06-18, then Viewpoint (evangelical) to c 23
567	529		RTE R1 (Tullamore 500)	412	see p40		V	0630-24; news in Irish 2030, also M-F 11,16,18, & $ 1115 & 18
576	520							
585	514		BBC R Solway (Dumfries 2)	(370)	94.7 93.1 94.1	250 0.64 0.31	M H V	M-F 0730-08, 1230-13; otherwise as R Scotland (370m)
594	505							

kHz	m	Station		MHz	kW		Remarks
603	498	Ind, Coast AM (Maidstone 0.1)	242				24h
		BBC R4 (Newcastle 2) Invicta FM:		102.8,97.0,95.9,96.1			(0.5 to 0.25 M)
		BBC R Gloucestershire (Gloucester 0.1)		95.0	0.1	M	From 06(S S̸ 07). S&W Reg 1805(S1930 S̸21)
				& 104.7	2	M	to 24 then R2. S̸ 17-19 R1.
							S 18-1845 Asian, 1845-1930 Afro-Caribbean.
612	490	<u>RTE R2</u> (Athlone 100)	235	see p40		V	24h
621	483						
630	476	BBC R Cornwall (Redruth 2)	457	103.9	17	M	From 06 (S 0630 S̸ 0730). S&W Reg M-S 1805--24 then R2. S̸ 17-19 R1, then R2.
		BBC R Bedfordshire (Luton 0.2)	258	103.8	0.5	M	24h inc R2, see 258m
639	468						
648	463	<u>BBC World Service</u> (Orfordness 500)		—			24h, inc "BBC 648": 0445-0545 G, 0630-07 Fr, 1130-12 Fr, 1630-17 G, 1730-1814 Fr, 1830-20 G. One hour later in 'October period' (see p9).
657	457	BBC R Clwyd (Wrexham 2)	<u>M-F</u>	-	-		07-0730,08-0830,09-1030,1655-17 | otherwise
-	-	BBC R Gwent: Newport	<u>M-F</u>	95.9	0.5	M	0710-0730,0810-10,1655-17 | R Wales
		Blaenavon	"	95.1	0.01	V	" " | (see 340m)
		BBC R Cornwall (Bodmin 0.5)	476	95.2	4.3	M	(Caradon Hill) See 476m
				& 96.0	0.06	M	(Scilly Isles)
666	450	Ind, DevonAir R (Exeter 0.34)	314	97.0	1	M	24h
		South West One (for E Devon):		103.0			(Stockland Hill)
		BBC R York (0.8)	238	103.7	2	V	From 06 (S S̸ 07). NE Reg 18-24 then R2.
				& 104.3	0.5	V	
675	444						
684	439						
693	433	<u>BBC Radio 2</u> (Droitwich 150; <u>Stagshaw</u>, Start Point, Westerglen 50, Burghead 25, Norwich 10, &c)	330	88 to 90.2		M	24h
702	427						

kHz	m	Station	MW	MHz	kW		Notes
711	422						
720	417	BBC R4 (London 0.75, Lisnagarvey 10, Londonderry 0.25)					
729	413	RTE R1 (Cork 10)	529	see p40		V	0630-24 then R2 to 0150. Irish news as 529m. Local pr M-F 1207-1330 & 17-1820.
		BBC Essex (Colchester 0.2)	393,196	103.5	12	M	From 06 (S Ꞩ 0630). Ꞩ-F 22-24 S&E Reg, S 1930-02 R1, then R2. Ꞩ 17-19 R1.
738	406	BBC Hereford & Worcester (Worcester 0.037)	366	104.0	2	M	From 06(S Ꞩ 07). Midland Reg M-F 21-24 then R2; R2 from 1930 S,19 Ꞩ. Ꞩ 17-19 R1.
747	401						
756	397	BBC R4 (Redruth 2)					
		BBC R Cumbria (Carlisle 1)	206,358	95.6	15	M	From 06(S 07 Ꞩ 0730). NW Reg 1930-24 then R2. Jazz F 2005-22.
		BBC R Shropshire (Shrewsbury 0.63)	189	96.0	4.8	M	From 06(Ꞩ 07). Midland Reg M-F 21-24 then R2. R1 S 1930, Ꞩ 1630, to 02 then R2. W 19-20 Classical, 20-21 Jazz.
765	393	BBC Essex (Chelmsford 0.5)	413,196	103.5	12	M	See 413m (add Classical Ŧ 19-20)
774	388	BBC R4 (Plymouth 1, Enniskillen 1)					
		BBC R Leeds (0.5)		92.4	5.6	M	From 06(S Ꞩ 07). NE Reg 18-24 then R2.
				& 95.3	0.04	M	Asian Ꞩ 19-2030 MW, Hindustani & Bengali.
		BBC R Kent (Canterbury 0.7)	290,187	104.2	10	M	From 06(S 0630 Ꞩ 07). M-F 22-24 S & E Reg. R2 from 24. Asian S 18-19 & 21-22. Ꞩ 18-19 Jazz. Ŧ 19-20 Folk.
		Ind, Severn Sound (Gloucester 0.14)		102.4	2	M	24h
			Stroud	103.0	0.1	M	
783	383						
792	378	BBC R Foyle (Londonderry 1)	<u>M-F</u>	93.1	31	M	Own progs 0730,0830-09,13-17 inc schs; otherwise as R Ulster, see 222m
		Ind, Chiltern R (Bedford 0.2)	193,362	96.9	0.9	M	24h

kHz	m	Station	MW also	VHF MHz	kW	Pol	Notes
801	374	BBC R Devon (Barnstaple 2)	351, 303,206	94.8	0.68	M	From 0605 (S̸ to 24). S & W Reg M-F 1805(S 1930)-24. R2 from 24.
810	374	BBC R Scotland (Westerglen 100, Burghead 100, Redmoss 5)	514,303	(Not S̸) 92.5 to 94.6; and (NW) 97.7 to 99.3 carrying the MW prog			0555 (S̸ 0630) - 0045. R4 till 0610 (M 0625 S 0655), S̸ 0740 + 08-0810. R4 from S 1945, S̸ 1530, M-F 24. VHF opt-outs M-F: R Tweed 0730-08,1730-18. R Orkney & R Shetland 0730-08,1330-14. R Shetland 1730-18 & 1815-1930(F 20). R Highland 0645-07,0730-08,1330-14,17-1730. In Gaelic, certain freqs 0810-09(T 10) 12-1230,1730-18, F 2210(T 2130)-2330; all freqs 1815-1930(MF 19) also S̸ 1530-16.
819	366	BBC R Hereford & Worcester (Hereford 0.037)		94.7	2	M	See 406m
828	362	BBC R WM (Wolverhampton 0.2)	206	-	-		WM = West Midlands. From 06(S 07 S̸ 0730). Midland Reg M-F 21-24. R2 from 24 (S S̸ 22). M-F MW 08-12 "WM Heartlands" (ethnic neighbourhood radio serving E Birmingham).
		Ind, Chiltern R (Luton 0.2)	378,193	97.6	1	M	24h
		Ind, 2CR (Bournemouth 0.27)		97.2	1	M	24h. 2CR = Two Counties Radio,
		Ind, R Aire (Leeds 0.12)		96.3	0.6	M	24h
		RTE: R na Gaeltachta (Corca 1)	See 311m				
837	358	BBC R Leicester (0.45)		95.1 & 104.9	0.3 / 8	M / M	From 06(S 0645 S̸ 07). R2 from 24(22 S, 18 S̸). Split freqs M-F 1805-24: Midland Reg M-F VHF 21-24. Ethnic prs MW T‡ 2030-2130. ‡ VHF 19-21 Jazz.
		BBC R Furness (Barrow 1.5)	Morecambe Bay:	96.1	3.2	M	Own pr M-F 0730-09, S 1005-1105; otherwise R Cumbria, see 397m
			Kendal:	95.2	0.1	M	
			Windermere:	104.2	0.06	M	

846	355					
855	351	BBC R Lancashire (Preston 1.5)	192	95.5 103.9	1.6 M 1.8 M	From 06 (S̸ 07). NW Reg 1930-24 then R2. W F S̸ 1830-1930 Asian.
		BBC R Norfolk (Norwich 1.5)	344	95.1	5.7 M	From 06. S & E Reg 22-24. R 1 00-02 (S 20-02) then R2.
		BBC R Devon (Plymouth 1)	374,303,206	103.4	10 M	From 0605 (to 24 S̸). S & W Reg 1805 M-F (S 1930) - 24. R2 from 24. Split: MW, Plymouth M-F 0835-10,1205-13; S 1315-14,1805-1930; S̸ 1205-14.
864	347					
873	344	BBC R Ulster (Enniskillen 1)	224			
		BBC R Norfolk (King's Lynn 0.3)	351	104.4	4.2 M	
882	340	BBC R Wales (Washford 100, Penmon 10, Tywyn 5)	457,267	-	-	From 0625 (S 07 S̸ 0730) inc R4 18-1830, 1905-1920, T F to 2145, Ŧ 20-2145, S 1825-1910. R2 M 2015-2045. R Cymru S̸ 1630-17. R2 from 22(MT 23,S 1930,S̸ 18). Welsh courses M-F 2145-22 & S̸ 17-1740. Asian M 1920-2015.
-	-	R Cymru:		92.4 to 94.6 (& Wenvoe 96.8)		0625(S S̸ 0755) - 0030 only. Inc R4 MTŦ 1905-22,W2045-22,S-Ŧ 2330-0030,S̸ 18-19; R Wales M-Ŧ 1830-1905; otherwise own progs in Welsh.
891	337	Private, TTT Radio (Dublin)		99.5		24h

900	333						
909	330	BBC Radio 2 (Moorside Edge 200, Brookmans Park 150, Westerglen 50, Clevedon 50, Lisnagarvey 10, &c)	& 433	88 to 90.2			
918	327						
927	324						
936	321	Ind, Brunel R (Chippenham 0.2)	258				
		GWR:		102.2	0.5	M	24h
945	317	Ind, Gem AM (Derby 0.2)	300				
		Radio Trent:		102.8	0.6	M	24h
954	314	Ind, DevonAir R (Torbay 0.32)	450	96.4	0.25	M	24h
		Ind, R Wyvern (Hereford 0.16)	196	97.6	0.78	M	24h
963	311	RTE: R na Gaeltachta	556,362	94.4	160	V	In Irish 08(S § 11) - 1330(§ 13) & 17-20.
		(Tir Chonail 10)		92.6,94.9	80,60	V	Otherwise VHF has RTE1.
				93.2,93.5	50,40	V	
				102.7,92.9	25,10	V	
972	309						
981	306						
990	303	Ind, Classic Gold (Doncaster 0.25)	230,194				
		Hallam FM:		103.4	1.5	M	24h
		BBC R2 (Tywyn 1)					
		BBC R Aberdeen (1)		93.1	60	H	M-F 0655-07,0730-08,12(WF 1230)-13, 1730-18, otherwise R Scotland (370m).
		BBC R Devon (Exeter 1)	374,351,206	95.8	0.4	M	
		Okehampton:		96.0	0.07	M	
		Ind, WABC (Wolverhampton 0.1)					"Nice & Easy"
		Beacon R: Wolverhampton		97.2	2	M	
		Shrewsbury(The Wrekin)		103.1		M	24h

999	300	BBC R Solent (Southampton 1)	221	96.1	5	H	From 06(S Ꞩ 0630), Ꞩ to 24, & ‡ F 21-24. S & W Reg 1805-21 ‡F (24 MTWS). R2 from 24. Ꞩ 17-19 R1. Split freqs Ꞩ 1445-17:MW sport.
		Ind, Gem AM (Nottingham 0.25)	317				
		R Trent:		96.2	0.3	M	24h
		Ind, Red Rose R (Preston 0.8)		97.4	1.9	M	24h
1008	298						
1017	295						
1026	292	BBC R Cambridgeshire(Cambridge 0.5)	207	96.0	1	M	Classical S 0830, W 20. From 06. M-F 22-24 S & E Reg. R2 from 24. Ꞩ 17-19 R1. 19-2030 M Folk, ‡ Eastern.
		BBC R Jersey (St Helier 1)		88.8	3.8	M	M-F 07-1805) M-S S & W Reg 1805-24 S 08-1330 then R2) then R2. Ꞩ 08-13 then R2; R1 19-02 then R2. Ꞩ Portuguese 0745-0810, Jersey Fr 0850-1920.
		Ind, Downtown R (Belfast 1.7)		97.4	1	M	05 (S Ꞩ 06) - 02.
		Londonderry		102.4	10	M	
		Limavady		96.4	2	M	
		Omagh & Enniskillen (Brougher Mtn)		96.6	10	V	
1035	290	BBC R Kent (Gillingham 0.5)	388,187	96.7	8.7	M	24h inc R2; see 388m
		BBC R Sheffield (1)		104.1	4.4	M	From 06(Ꞩ 07) inc Ꞩ 21-23. Asian S 18-22 MW,
				88.6	0.32	M	ethnic Ꞩ 18-21. Otherwise NE Reg 18-24 then R2
		Ind, NorthSound R (Aberdeen 0.78)		96.9	0.6	M	06 - 24.
		Ind, West Sound (Ayr 0.32)		96.7	0.8	M	06 (S Ꞩ 07) - 20 (F S Ꞩ 21)
		Girvan		97.5	0.15	M	
1044	287						
1053	285	BBC Radio 1 (Droitwich 150, Start Point 100, Stagshaw 50, Burghead 20, Norwich 10, &c)	275, 271, 202	97.7 to 99.5			05 - 02 . Top 40 (Ꞩ 17-19) also broadcast on all VHF of R2 & on many BBC local stns.
1062	282						
1071	280						

kHz	m	Station	MHz	kW		Hours
1080	278					
1089	275	BBC Radio 1 (Brookmans Park, Moorside Edge 150, Washford, Westerglen 50, Lisnagarvey 10, &c)	97.7 to 99.5			05 - 02. See 285m.
1098	273					
1107	271	BBC R1 (Merseyside 0.5)				
		BBC R Northampton (0.5)	104.2	4	M	From 06(S 0645 Ꞩ 07). Ꞩ - Ŧ 22-24 SE Reg.
		Geddington	103.6	0.8	M	R1 00-02 then R2.
		Ind, Moray Firth R (Inverness 1.5)	97.4	2.8	M	M-F 06-14, S 06-02, Ꞩ 07-24.
1116	269	BBC R Derby (1.2)	94.2	0.01	V	From 0630 (S 07 Ꞩ 08). Midland Reg M-F 21-24 then R2; R2 from 1830 S, 21 Ꞩ. 19-21 M Folk, Ŧ Jazz. Asian F 1830-21. Ꞩ 14 R2, 17-19 R1.
		Sutton Coldfield	104.5	5.4	M	
		Stanton Moor	95.3	1.2	M	
		BBC R Guernsey (St Peter Port 0.5)	93.2	1	M	M-F 07-1805) M-S S&W Reg 1805-24 S 08-1330 then R2) then R2. Ꞩ 08-14 then R2; R1 17-02 then R2.
1125	267	BBC R Wales (Llandrindod Wells 1) 340				
1134	265					
1143	263					
1152	261	Ind, London Talkback R (23.5)				24h. LBC's major discussions are now on VHF.
		Crown FM:	97.3	4	M	
		Ind, R Clyde (Glasgow 3.6)				
		Clyde FM:	102.5	20	M	24h
		Ind, Xtra AM (Birmingham 3) & 1359				
		BRMB R:	96.4	2	M	24h
		Ind, Piccadilly R (Manchester 1.5)				
		Key 103:	103.0	4	M	24h
		Ind, Great Northern R (Newcastle/Tyne 1.8)				24h
		Metro FM:	97.1	10	M	also 103.0

kHz	m	Station		VHF	kW		Notes
1152 (cont.)	261	Ind, Plymouth Sound (Plymouth 0.32)		97.0	1	M	24h
		Tavistock:		96.6	(0.04);		"R in Tavistock" 5½ hrs M-F, 3h S, 2h S̸.
		Ind, R Broadland (Great Yarmouth 0.8)		102.4	3.3	M	24h
1161	258	BBC R Sussex (Bexhill 1)	219,202				From 0530 (S S̸ 07). M-Ŧ 22-23 usu S&E Reg.
		Heathfield:		104.5	10	M	R2 T 1805-22, S 1830-22, S̸ 16-17 & 2030-22. S̸ 17-19 R1. Focus(local magazines) M W Ŧ 19-22. R2 from 24.
		BBC R Bedfordshire (Bedford 0.1)	476	95.5	1	M	From 06(S 07 S̸ 0635). S&E Reg 22-24 then R2.
				104.5	2.2	M	S̸ 17-19 R1. Asians M 21-22. Italians T 0835-10.
		Ind, Brunel R (Swindon 0.16)	238,321				24h
		GWR: Swindon		97.2	0.22	M	
		Marlborough		96.5			
		Ind, R Tay (Dundee 1.4)	189	102.8	1.3	M	24h
		Ind, Classic Gold (Hull 0.35)					24h
		Viking FM:		96.9	10	M	
1170	257	Ind, R Orwell (Ipswich 0.3)	240	97.1	4	M	0555 - 24.
		Ind, The Gold AM (Portsmouth 0.12)	193				
		Ocean Sound:		97.5	0.85	M	24h
		Ind, Swansea Sound (0.58)		96.4	1	M	24h
		Ind, Gt N R (Stockton-on-Tees 0.32) & 1152					24h
		TFM R:		96.6	8.9	M	
		Ind, Signal R (Stoke-on-Trent 0.2)		102.6	1.9	M	24h. VHF splits.
		Cheshire:		96.4			
		Stafford:		96.9			
1179	254						
1188	253						
1197	251	BBC Radio 3 (Bournemouth 0.5, Torbay 1, Cambridge 0.25, Enniskillen 1)					See 1215m
1206	249						

1215	247	<u>BBC Radio 3</u> (Moorside Edge 100, Washford 60, Brookmans Park 50, Westerglen 50, Droitwich 30, Burghead 20, Lisnagarvey 10, &c)	& 251	90.2 to 92.6 (Channel Is. 94.8)			0655-0005. VHF usu has Open University prs 0635-0655.
1224	245						
1233	243						
1242	242	Ind, Coast AM (Gillingham 0.32)	497				
		Invicta FM:		103.1	0.63	M	
1251	240	Ind, Saxon R (Bury St Edmunds 0.76)	257	96.4	2	M	0555 - 24 inc relays of R Orwell (257m)
1260	238	BBC R York (Scarborough 0.5)	450	95.5	0.25	M	24h, see 450m
		Ind, Brunel R (Bristol 1.6)	258,321				24h
		GWR:	Bristol	96.3	2	M	
			Bath	103.0	0.08	M	Own prog 06-09
		Ind, Marcher MF (Wrexham 0.64)					24h
		Marcher Sound:	Wrexham	103.4	1.4	M	
			Deeside	97.1			
		Ind, Gem AM (Leicester 0.3)	317				24h
		Leicester Sound:	103.2	103.2	0.4	M	
1269	236						
1278	235	RTE R2 (Dublin 10, Cork 10)	490	see p40			24h
		Ind, Classic Gold (Bradford 0.43)	196				
		Pennine FM:		97.5	0.5	M	24h
1287	233						
1296	231	BBC World Service (Orfordness 500). Mainly beamed to E Europe, in their languages. Silent spells; p21.					
1305	230	Ind, Red Dragon R (Newport 0.2)	220	97.4	0.5	M	M-F 15-1730, otherwise as 220m. 24h.
		Ind, R Hallam (Barnsley 0.15)	194,303	102.9	0.5	M	24h
1314	228						
1323	227	Ind, Southern Sound (Brighton 0.5)		103.5	0.9	M	24h. Also Newhaven, 96.9 (0.1).
		FM separate: Eastbourne area (Heathfield 102.4), Hastings 97.5.					

kHz	m	Station	MW	VHF	kW	Pol	Notes
1323 (cont.)	227	BBC R Bristol (Taunton 0.63)	see 194	95.5	9	M	MW this freq inc "Somerset Sound", M-F
			Bath:	104.6	0.82	M	0655-09, 1205-1310, 1630-18;S 08-10,Ꞩ 12-14.
1332	225	Ind, Hereward R (Peterborough 0.6)		102.7	4	M	24h
		BBC Wiltshire Sound (Chippenham 0.65)	219	104.3	0.6	M	24h, see 219m
1341	224	<u>BBC R Ulster</u> (Lisnagarvey 100)	378,344	93.0 to 94.5			From 0630 (S 08 Ꞩ 0855). Inc R4 M-F 18-1830, Ꞩ 18-1815. R2 (but on VHF R4) M-F 1920-23, S 18-23,Ꞩ 19-22; R2 nightly from 0005. VHF inc R4 educ prs & starts 0555 Ꞩ R4. In Irish: M-F 1902-1920, Ꞩ 1815-19. Music 23-24 M Country, T Folk, W Jazz, Ŧ Shows, F Classical, S Swing.
1350	220						
1359	220	BBC R Solent (Bournemouth 0.85)	300	-	-		24h inc R2, see 300m
		Ind, Xtra AM (Coventry 0.27)	261				24h
		Mercia Sound:		97.0	2	M	
		Ind, Red Dragon R (Cardiff 0.2)	230	103.2	0.2	M	24h
		Ind, Breeze AM (Chelmsford 0.28)	210				
		Essex Radio:		102.6	2	M	24h
1368	219	Manx Radio (Foxdale 20, Isle of Man)		97.3	10 (& 89.0 0.05)		Commercial;Manx govt owned; 07-01.
		BBC R Sussex (Crawley 0.5)	Reigate:	104.0	3.8	M	See 258m & 202m. 24h
		BBC R Lincolnshire (Lincoln 2)		94.9	6	M	From 0630(S 07 Ꞩ 0730). Midland Reg M-F 21-24 then R2. S 17 R2, 1930-02 R1; Ꞩ 17-02 R1;then R2. Ŧ 19-21 MW Jazz.
		BBC Wiltshire Sound (Swindon 0.4)	225	103.6	0.5	M	From 06 (S 0630 Ꞩ 07). S & W Reg 1805(Ꞩ 22)-24
			Salisbury	103.5	0.03	M	then R2. Ꞩ 1550-17 Classical.
1377 to 1422:		no U.K. stations. (See p38 for new Sunrise R on 1413.)					
1431	210	Ind, Breeze AM (Southend 0.35)	220				24h
		Essex Radio:		96.3	1.4	M	
		Ind, Radio 210 (Reading 0.14)		97.0	0.5	M	Also Basingstoke 102.9 (4 M). 24h.

kHz	m	Station		MHz	kW		Notes
1440	208	RT Luxembourg (1200). English pr 1845 ($ 1815) to 03. G 0430 (S 05 $ 07);Dutch 18($ 17);Italian 1830($ 18).					
1449	207	BBC R Cambridgeshire (Peterboro 0.15)	292	95.7	5.1	M	24h inc R2, see 292m
		BBC R4 (Redmoss 2)					
1458	206	BBC, Greater London R (GLR)(Brookmans Park 50)		94.9	4	M	From 0530. 22-24 usu S&E Reg, but S 'The Saturday Debate'. R1 00-02 then R2. Split freqs S 19-22 $ 19-24. S 19-21 VHF Jazz, S21-22 MW Asian.
		BBC R WM (West Midlands)(Birmingham 5)	362	95.6	11	M	From 06(S07 $0730).Reg M-F 21-24.R2 24(S$ 22).
		BBC R Devon (Torbay 2)	374,351,303				
		BBC R Newcastle (2)		104.4	0.042	V	From 06(S 0630 $ 0730). NE Reg 18-24 then R2.
				95.4	3.5	H	$ 16-17 Asian & Chinese. $ 17-19 R1.
		Chatton:		96.0	5.6	M	
		BBC Greater Manchester R (GMR)(5)		95.1	5.6	M	From 06(S $ 07). NW Reg 1930-24 then R2. T 1830-1930 Eastern;S 18-1930 African & Caribbean; $ 1830-1930 Asian.
		BBC R Cumbria (Whitehaven 0.5)	397,358	95.6	15	M	24h inc R2, see 397m
1467	204						
1476	203	Ind, County Sound Gold (Guildford 0.5)					
		County Sound:		96.4	3	M	24h . Also Haslemere 97.1 (Delta R).
1485	202	BBC R Merseyside (Liverpool 0.5)		95.8	3	M	From 06(S 0630 $ 07). NW Reg 1930-24. R2 from 24(F 02). M-F 1830-1930 Afro-Caribbean; S 1805-19 Asian, 19-1930 Chinese.
		BBC R Humberside (Hull 1)		95.9	9.6	M	From 06(S $ 07). NE Reg 18-24 then R2.
		BBC R Oxford (0.5)		95.2	5.8	M	From 06(S 07 $ 08). M-F 22-24 S & E Reg. R2 from 24 (S $ from 18). Classical M 2030.
		BBC R Sussex (Brighton 1)	258,219	95.3	1.2	M	24h inc R2, see 258m
		Newhaven:		*95.0	0.1	M	
		BBC R4 (Carlisle 1), and BBC R1 (Bournemouth 2).					
1494	201						

1503	200	BBC R Stoke-on-Trent (1)		94.6	6.1	M	From 06(S Ꞩ 07). Midland Reg M-F 21-24 then R2. R2 Ꞩfrom 1930 S, 24 Ꞩ. Ꞩ 17-19 R1. W 20-21 Jazz. Ŧ 19-21 Asian.
1512	198						
1521	197	Ind, R Mercury (Reigate 0.64)		102.7	3.6	M	24h
		Horsham:		97.5	0.03	M	
		BBC R Nottingham (0.5)	189	103.8	0.3(*1)	M	From 06(S Ꞩ 07). Midland Reg M-F 21-24 then R2; R2 from 18 S, 21 Ꞩ. M 19-21 Folk. Ꞩ 1530-19 Afro-Caribbean. Some splits e.g. Ŧ 19-21 MW young Asians, VHF Jazz; F 19-21 MW Asians.
1530	196	BBC Essex (Southend 0.15)	393,413	95.3	1.2	M	24h inc R2, see 413m
		Ind, R Wyvern (Worcester 0.5)	314	102.8	1	M	24h
		Ind, Classic Gold (Huddersfield 0.74)	235				24h
		Pennine FM:		102.5	0.63	M	
1539	195						
1548	194	Ind, Capital Gold (London 97.5)					24h. Tel 0898 400747 to listen in.
		Capital Radio:		95.8	4	M	24h.
		Ind, R City (Liverpool 4.4)					24h
		R City FM:		96.7	8.2	M	24h
		Ind, R Forth (Edinburgh 2.2)					24h
		R Forth FM:		97.3	9.8	M	
		Ind, Classic Gold (Sheffield 0.74)	230,302				24h
		Hallam FM:		97.4	0.4	M	
		Rotherham		96.1	0.05	M	
		BBC R Bristol (5)	227	94.9	0.95	M	From 0530(S 0645 Ꞩ 07). S & W Reg 1805(Ꞩ 22) to 24 then R2. Ꞩ 17-19 R1.
		BBC R Cleveland (Stockton-on-Tees 1)		95.0	5	H	From 06(S 07 Ꞩ 0730). Ꞩ 17-19 R1; F 1805-1930
		Whitby:		95.8	0.04	H (* 0.08 M)./	MW Asian; otherwise daily NE Reg 18-24 then R2.

kHz	m	Station		MHz	kW		Notes
1557	193	BBC R Lancashire (Lancaster 0.25)	351	104.5	2.1	M	24h inc R2, see 351m
		Ind, Northants 96 (Northampton 0.8)		96.6	4	M	24h. (Chiltern R.)
		Ind, The Gold AM (Southampton 0.5)	257				24h. (Ocean Sound.)
		The Power FM:		103.2	2	M	
		The Light FM (Winchester)		96.7	0.5	M	
1566, 1575 & 1593:		no U.K. stations.					
1584	189	BBC R Nottingham (Mansfield 1)	197	95.5	2	M	24h inc R2, see 197m
		BBC R Shropshire (Woofferton 0.5)	397	95.0	0.01	M	24h inc R2, see 397m
		Ind, R Tay (Perth 0.21)	258	96.4	0.35	M	24h
1602	187	BBC R Kent (Tunbridge Wells 0.25)	388,290	-	-		24h inc R2, see 388m.

Local stations on VHF/FM only:-

BBC CWR (Coventry-Warwickshire Radio) on 94.8 for Coventry & N Warwickshire & 103.7 For Stratford & S Warwickshire.
(Ind) CN FM (Cambridge & Newmarket) 103.0.
(Ind) Fox FM: Oxford 102.6, Banbury 97.4.
(Ind) Horizon R: Milton Keynes 103.3.
(Ind) Orchard FM (for Yeovil & Taunton): so far Mendip is on air, 102.6.
(Ind) R Borders (inc Berwick) on 97.5, 96.8, 103.1, 103.4, all V.
Planned (not included in our VHF list):
(Ind) South West Sound (Scotland) ? May 90.
BBC R Suffolk on 104.6 & 103.9, ? Apl 90.
BBC R Dorset on 103.8 & 95.7, ? Apl 91.
BBC R Surrey & Berkshire, ? Apl 91, on 104.6,104.4,104.1,94.6,95.4.

Community Radio stations now holding franchises:-

VHF/FM only:- Greater London: London Jazz R (LJR); Kiss FM; Melody R. Haringey (ethnic): shared by London Greek R & WNK R (103.1, on air). Brixton (ethnic): South London R. Thamesmead: RTM. Bristol: FTP. Manchester (ethnic): Sunset R (102.0 V, on air). Birmingham: Buzz FM (*102.4). Stockport (Gtr Manchester): KFM R (104.9 V, on air). Coventry (ethnic): R Harmony. Bradford (ethnic): Bradford City R. Sunderland Community R, Belfast C.R., Stirling C.R. Glasgow (Easterhouse): East End R.

MW only:- Greater London (ethnic): Spectrum R. Hounslow (ethnic): Sunrise R (1413 kHz, on air). Heathrow & Gatwick: Airport Information Radio. Tendring: NE Essex C.R. Isle of Wight R. Kettering (Northants): KNBC R. Bathgate, W Lothian (Scotland): TBA.

If any independent station cares to send me programme information or new data at any time, I shall be most grateful.

G.W. Jan.1990.

VHF

A page for your own pencil notes. The band has been allocated to broadcasting up to 104 MHz so far, and there is very little beyond 104.9. For present frequencies consult checklist on next pages, & pp 25-38. All VHF transmitters are now stereo or soon will be.

88	**98**
89	**99**
90	**100**
91	**101**
92	**102**
93	**103**
94	**104**
95	**105**
96	**106**
97	**107**
	108

VHF (FM)

The following list includes every frequency for U.K. local radio, correct to January 1990. Square brackets & arrows show the present frequency ranges of BBC national & regional programmes; full details of all relay stations can be had from BBC, see p8. Other countries' networks are scattered throughout the band, so only a few are entered here. The conventional channel numbers at left are sometimes met with. Sd = Sound. RF = Radio France. F Mus = France Musique. FC = France Culture. F I = France Inter. RNG = Radio na Gaeltachta in Irish + RTE 1 & 3 in English.

Ch	MHz	Station	MHz	Station	MHz	Station
3	87.8	Hilversum 2, Goes	87.9		88.0	
4	88.1	[Radio 2 →	88.2	RTE1	88.3	
5	88.4	FC Amiens	88.5		88.6	R Sheffield
6	88.7	F Mus Lille	88.8	R Jersey/RTE1	88.9	
7	89.0	Manx R	89.1	RTE1	89.2	FC Cherbourg
8	89.3		89.4	F Mus (Brest, Boulogne)	89.5	
9	89.6	RTE1	89.7		89.8	F Mus Abbeville
10	89.9	F Mus Rennes	90.0	RTE1	90.1	
11	90.2	← R2] F Mus Neufchatel	90.3	[Radio 3 →	90.4	RTE2/BRT3
12	90.5		90.6		90.7	
13	90.8		90.9		91.0	RTE2
14	91.1	R2 Channel Is.	91.2	Manx R	91.3	RTE2
15	91.4		91.5	FC Caen	91.6	
16	91.7		91.8		91.9	
17	92.0	F Mus Rouen	92.1		92.2	RTE2
18	92.3	F Mus Cherbourg	92.4	← R3] R Leeds	92.5	R Highland/ [R4, R Scotland, R Cymru →
19	92.6	R Highland/RNG	92.7	R Aberdeen, Orkney, Shetland/ F I Neufchatel	92.8	R Tweed
20	92.9	R Highland/R nan Eilean/RNG	93.0	RF Bretagne Ouest(Brest)/ RF Picardie(Amiens) [R Ulster →	93.1	R Aberdeen/Solway/Foyle/ F I Abbeville
21	93.2	R Guernsey/RNG	93.3	R Highland/FC Le Havre	93.4	
22	93.5	R Highland/R nan Eilean/ R Tweed/RNG/ F I Rennes	93.6		93.7	R Aberdeen/Orkney/ Highland

Ch			
23	93.8 R Aberdeen	93.9 R Tweed/R nan Eilean/Highland	94.0 FC Rouen/R Highland
24	94.1 F I Cherbourg/R Solway/Highland	94.2 R Derby	94.3
25	94.4 RNG	94.5 ← R Cymru] R Aberdeen/Highland	94.6 ← R4; R Scotland] R Stoke/Highland
26	94.7 RF Frequence Nord(Lille)/RNG/ R Hereford & Worcester/Solway	94.8 R Devon/CWR/R3 Channel Is.	94.9 GLR/R Lincolnshire/ Bristol/RNG
27	95.0 R Cleveland/shropshire/ Gloucestershire/*Sussex	95.1 R Gwent/Leicester/Norfolk/ GMR	95.2 R Cornwall/Oxford/Furness
28	95.3 R Sussex/Leeds/Derby/BBC Essex	95.4 ← R Ulster] R Newcastle/ F I Brest	95.5 Freq Nord Boulogne/R York/ Lancs/Bedfds/Notts/Bristol
29	95.6 R WM/Cumbria/F Mus Caen	95.7 BRT1(Egem)/R Cambridgeshire	95.8 R Merseyside/Cleveland/ Devon/Capital R
30	95.9 R Humberside/Gwent/Invicta R	96.0 R Newcastle/Cambridgeshire/ /Shropshire/Cornwall (Scilly Is.) /Devon/R1,Belfast/FC Neufchatel	96.1 R Furness/Solent/Hallam / Invicta R
31	96.2 R Trent	96.3 R Aire/GWR/Essex R	96.4 DevonAir R/Downtown R/R Tay/ Swansea Sd/County Sd/ Saxon R/BRMB R/Signal R
32	96.5 GWR/ RF Normandie (F I) Rouen	96.6 Downtown R/Plymouth Sd/ TFM R/Northants	96.7 R Kent/City/Ocean Sd/ West Sd
33	96.8 R Cymru, Wenvoe/R Borders/ Hilversum 3, Lopik	96.9 Southern Sd/Chiltern R/ Signal R/Viking/North Sd	97.0 Invicta R/DevonAir R/ R 210/ Mercia Sd/Plymouth Sd
34	97.1 R4 Ch Is/Metro/Orwell/Marcher	97.2 GWR/R Bedfds/2CR/Beacon R	97.3 Crown (LBC)/R Forth/Manx R
35	97.4 FC Abbeville/Downtown R/Fox/ Hallam/Red Rose R/ Red Dragon R/Moray Firth R	97.5 Ocean Sd/Southern Sd/R Mercury Pennine/R Borders/West Sd	97.6 R Wyvern/Chiltern R
36	97.7 [R 1; R Scotland NW area →	97.8 FC Brest	97.9
37	98.0 FC Lille	98.1	98.2
38	98.3 FC Rennes	98.4	98.5 F Mus Le Havre
39	98.6 BRT2, Egem	98.7	98.8
40	98.9	99.0	99.1
41	99.2	99.3 ← R Scotland NW area]	99.4 F Mus Amiens

Ch	MHz	Station	MHz	Station	MHz	Station
Ch 42	99.5	← R 1] TTT R, Dublin	99.6	F I Caen	99.7	
43	99.8		99.9	FC Boulogne	100.0	
44	100.1	BRT2 + Rg(Egem)	100.2		100.3	
45	100.4		100.5	Sunshine R (Dublin)	100.6	RF Picardie Abbeville/BRT2,
46	100.7		100.8		100.9	/Veltem
47	101.0	Sunshine R (Dublin)	101.1	R Classique (Paris)	101.2	
48	101.3		101.4		101.5	Sunshine R (Dublin)
49	101.6		101.7		101.8	
50	101.9		102.0	Sunset R (Manchester)	102.1	
51	102.2	GWR	102.3		102.4	Severn Sd/Southern Sd/ R Broadland/Downtown/*Buzz
52	102.5	R Clyde/Pennine R	102.6	Essex R/Fox/Orchard/Signal	102.7	R Mercury/Hereward R/RNG
53	102.8	R Tay/Wyvern/Trent/Invicta	102.9	R 210/Hallam	103.0	Key 103, Manchester/CN FM/ GWR/Severn Sd/South West 1
54	103.1	Beacon/Invicta/R Borders/ London Greek R & WNK R/ RF Armorique (F I) Rennes	103.2	Ocean Sd/Leicester Sd/ Red Dragon R	103.3	Horizon R
55	103.4	R Devon/Hallam/Borders/ Marcher Sd	103.5	BBC Essex/Wiltshire Sd/ Southern Sd	103.6	R Northampton/Wiltshire Sd
56	103.7	R York/CWR/F I Lille	103.8	R Nottingham/Bedfordshire	103.9	R Cornwall/Lancashire
57	104.0	R Sussex/Hereford & Worcester /R Cymru, Blaenavon	104.1	R Sheffield	104.2	R Kent/Furness/Northampton
58	104.3	R York/Wiltshire Sd	104.4	R Newcastle/Norfolk	104.5	R Derby/Sussex/Lancashire/ Bedfordshire
59	104.6	R Bristol	104.7	R Gloucestershire	104.8	R 1 (London) at present
60	104.9	R Leicester/KFM R, Stockport	105.0		105.1	

PROGRAMME NOTES

BBC Radio 4 has its full home service only on longwave (with the excellent World Service filling out the 24h from the Droitwich transmitter only). VHF carries educational progs in school hours, also S & $ 16 - 18 ('Options', inc usu a language course at 17 and/or 1730). VHF 0030 - 0110 T - S in term-time has more schools broadcasts, intended for tape-recording by time-switch. LW has shipping forecast at 0033, also 0555, 1355 & 1750. - With the main news once more at 22, the hour before midnight has become quite enterprising & varied. It may include humour, musical or historical features, satire (e.g. Week Ending), as well as Today in Parliament (usu 2330).

BBC Radio 3 is nearly 90% music, mostly concert music. The 20th century has been given more time recently (esp what many people would class as modern & experimental — some of it actually 80 years old!) But the romantic 19th & the classical & baroque 18th are still very well represented. About 12% of time is devoted to the spoken word: drama, & more talks & discussions than two years ago. Rarer elements are poetry, jazz & cabaret, ethnic music, early music: some weeks hardly any, then a short season. — VHF has the full service (+ usu Open University S 0635 - 0655); MW still includes ball-by-ball cricket this season.

BBC Radio 2 (varied light entertainment) 24h; some sports commentaries on MW.
Radio 1 (the pop programme) now broadcasts 05 - 02 & has its own VHF band.
Brief news on the hour (R2), on most half-hours (R1).
All BBC locals also relay much R2, esp at night (see pp 25 - 38).
Radio 5 is due on MW from 27 August 1990; see p2.

Switzerland: On 765 kHz MW, every night from 2305 to 0500, programme of mainly classical music; some operetta; some avant-garde; occ jazz 23 - 01. Identified near each hour as "Notturno: DRS Zwei; Espace Deux; Rete Due". See pp 4-5. Briefest announcements, in German or French or Italian. — Espace 2 is also relayed on SW 08 - 11.

Radio France, medium wave (Ondes Moyennes, or Petites Ondes):

0530 M-S RFI, for immigrant communities: M Vietnamese & Turkish, T Laotian & Serbo-Croat, W & Ŧ French & Spanish, F Kampuchean & Spanish, S Turkish.

06 - 18 Radio Bleue. [Stop Press, Feb. '90:] "Varied & grown-up & lively; strongly recommended," as we said last time; and it has just won the extension right through the day. This programme is truest to the real France: a sensible concern for the present, with loving regard to the past, & esp France's hundred years of so of great popular songs & ballads: the chanteurs, chanteuses, chansonniers, through to those of today. Memory lane, rather than golden oldies. Comedians too, & wireless archives. (When did we last hear a recording of John McCormack, or Harry Tate?) Much phoning-in.

11 - 13 S R France Armorique keeps its opt-out programme in the Breton language (Rennes 711, Brest 1404).

18 - 1815 France Inter main news; then some transmitters fill in with continuous heavy pop, with one comedy slot (no I.D.) till the RFI programmes.

18 - 1830	M - F	Lille 1377 only: Polish.	Radio France Internationale
18 - 19	M - F	Strasbourg 1278 only: German.	
2030 - 21	S	For Yugoslavs.	
21 - 2130	M - S	Portuguese (& French).	
2130 - 22	M - S	Arabic (& French) for Maghrébins (N Africans).	
22 - 23		Spanish, on Toulouse 945 only.	

France Musique (VHF 24h, closed W 00 - 05): classical music, avant-garde, ethnic, jazz, festival reports, interviews &c. Jazz 1107-1130 M-F, 1630-17 M-Ŧ, S 13-14, $ 18-19.
France Culture: VHF 24h, closed W 00 - 06.
France Info: factual magazine programme (main cities, around 105 - 106 MHz VHF).
France Inter: the main home service, 24h (closed LW & MW T 00 - 04, VHF W 00 - 05).

Spain: RNE1 stations are all between 585 & 855 kHz; there are also a large number of local stations, mainly organised under the networks SER, RCE or COPE. Programmes are mostly speech. RNE2 (music & cultural) & RNE3 ('young people's music') are on VHF only, as are broadcasts in Catalan (Barcelona) and Galician (La Coruña).

The VHF band is worth exploring, wherever you are in Europe. Besides giving the best reception, and some quality programmes not receivable here, it will include private stations, and local programmes useful for tourists and/or travellers, occ in English. Note Elektra 108, giving BBC World Service on the French & Italian Riviera.

CLASSICAL MUSIC AND JAZZ

(See also British locals on pp 25-38; and on VHF try France Musique, 24h.)

Morning

DLF. BBC R3 esp c 1130-14. R2 $ 0905-11. BBC W/S GMT 0815 $ F, 0945 T, 1030 F (W jazz). Poland SW (49mb 41mb) 1130-12 Chopin.

Afternoon

BRT3. DLF. BBC R3. FIP, all sorts, to c 16. BBC W/S GMT Ŧ F $ 1515; Ŧ 1345 jazz or folk. Austria ORF1 16. Vatican 16.

Early evening

BBC R3 M-Ŧ 1730-19; jazz F1745 usu & S 17-1745. Austria ORF2 16 & 18. Try Bulgaria c 1845, folk. Italy 2 18-19 & occ Italy 1. 19 +, DLF, Poland 2. BRT1 M 19 jazz. BBC R2 W 19-2030 folk.

Mid-evening

BBC R3 $ M W Ŧ c 1930. R2 F S 1930, $ 21; M jazz 21. R4 S 2115. W/S GMT $ 2115, W 2145. RIAS 20+. c 19 try Norway, Hungary. Jazz: BRT1 Ŧ 20+, RTBF1 21-22, Hilversum 1 20+ & try H2 & 3. France Inter S $ 20-2130. Luxembourg LW $ 20-21. c 20, DLF, Czech, Poland 1 & 2. 21+, Austria ORF1, Italy 1 & 2. Suisse(Sottens) M-S 19-2130.

Late evening

R Suisse (Sottens) M-F 2140-23. From c 22, Bucharest 2, Poland 2, Czech Praha, Bratislava, RIAS, ORF2, Ljubljana. Denmark 3 jazz. BBC R2 $ 23-01 jazz. R Moscow W/S Ŧ 2230-23. BBC W/S $ 2115, W 2145; T Ŧ 2315.

BBC R3 23-24 usu, & alternate S jazz. From c 23, Vatican, Czech Hvezda, DLF Nachtkonzert, occ Yugoslavia, all to 24. Poland 2 (& 1503). Prague Inter (1287 &c) to 0030.

Early hours

R Suisse (Sottens) 2305 - 0500. DDR 00-0030. BBC W/S GMT $ F 0030, S 0045, M 0145; F 0130 jazz or folk. W German stations esp at 02. Bulgaria c 01-0130. Italy esp 0130-02, 0330-04. Czech LW 02. R Moscow 02. Austria 03 0530.

One starting point of my dial-searching was to find some classical music after 23 hours, when Radio 3 used to close down. Most composers and many nations have an original voice which a sensitive listener learns to spot and appreciate after a few bars. European radio is a splendid random source of music, ideal for the solo games of "Who wrote that?" or "What comes next?" Some of it is familiar and welcome; some things are familiar and no longer welcome... and every so often you find a piece that is new to you and so good that you add it to your own personal heritage — and perhaps to your record collection. Without Swiss radio, I would never have known the brilliant musical jokes of Franz Tischhauser, as hilarious as the Hoffnung Festivals or Dudley Moore at the piano.

Seemingly much more limited by its structure, jazz too has an extraordinarily wide range: musicality & originality are essential to it. I love its freshness; the more so because this is one type of music I have never played or written. Another deep love of mine is Indian music, one of the richest traditions in the world; it is time it had a regular place on our national radio again.

The notion that one ought not to have music just as a background, especially in the home, is ludicrous. One has pictures on the wall, but is not looking intently at them all the time. Radio can be followed in the mind, though: its words and thought; its music and the thought in the music; while we do other things. Or rather the radio follows <u>us</u>.

One has pictures; one also has wallpaper... Twiddle the dial, and what do you find, usually? I listen, as we all do, but after ten minutes I am unable to take the constant four-in-a-bar, always so loud and goey or gooey, the stereotyped harmonisation in a modern 'organum' of consecutive fifths, the inability to close without fading. The odd original piece stands out a mile; I might even buy the single or the sheet music; but that's rare.

See if you can define, in musical terms, the originality of the TV theme from The Bill; or of the "Classical" Symphony of Prokofiev...

I quite like folk, and country, and music from the shows; I enjoy cabaret songs (often songs of protest) in their perfectly articulated poetic French or German; my favourite request programme is Anything Goes (BBC World Service); but what my readers mostly appreciate and expect is guidance towards jazz and classical music (lots of which becomes middle-of-the-road). And at present that is what, in both senses, I have time for.

G.W.

BROADCASTS IN ENGLISH

This is a selection of English broadcasts, mostly beamed to Europe, which are generally heard well. The times & frequencies listed here cannot be completely up-to-date for long: broadcasters make changes to their schedules from time to time, and you can expect many SW frequencies to be altered early in March, May, September & November, owing to seasonal changes in the ionosphere. It is a good plan to listen to opening & closing announcements, and to "DX programmes", keeping a notebook or cassette-recorder handy. Make notes on this page.
Times, this page: GMT (UTC). dx = This transmission includes a DX programme on ...days.

06 Vatican to 0620, Norway $ -0630, RBI (R Berlin International, East Germany) -0645, WYFR -07, WCSN -08.
0615 Canada M-F -0630.
0630 Albania, Poland -07. AWR Europe -0730.
0645 Canada M-F -07.

07 Japan -08, HCJB Ecuador -0830, Australia -10.
0730 Finland -0755, Bulgaria, Swiss -08.
0740 TWR M-F -0925 S -0955 $ -1045.
0745 Czech -08.

08 South Korea -09.
0830 Austria -09.
0840 Greece M-S -0850.
0845 Czech -09, RBI -0930.

09 AWR Europe $ dx -10.
0945 Czech -10.

1030 Dubai -11 dx S $, AWR Eur -1130.
1045 Czech -11, RBI -1130.

11 Pakistan -1120, Israel -1130, Deutsche Welle (Germany) -1150, South Africa -12.
1130 Netherlands -1225 dx Ŧ, Austria -12.
1145 Czech -12.

12 Poland -1225, RBI -1245.
1230 France -13.

13 Romania, Swiss dx S, Norway $ -1330, N Korea -1350.
1330 Dubai -14.

14 RBI -1445, Malta -16, South Africa, WCSN -16.
1430 Neths -1525 dx Ŧ, Austria -15, Poland dx W -15.
1445 Vatican -15, RBI -1530.

15 Japan dx $ -16, Jordan -1730.
1505 Finland -1530.
1530 Sweden dx T -16, Czech -1625, Australia -2030.
1545 Canada -16.

16 Poland dx T -1630, France -17, Dubai -1645, WYFR -17, WHRI -21.
1615 Hungary dx M Ŧ -1630.

1715 Canada -1730.
1718 Pakistan -18.
1730 Austria -18.

18 Israel -1815, Czech, Vietnam, Sweden dx T -1830, S Africa dx $, Brazil -19, Kuwait -21.
1815 RBI -19.
1830 Poland dx W, Swiss, Belgium, Albania -19, Neths dx Ŧ -1920, Afghanistan -1930.
1845 India -2230.

19 Norway $ -1930, Spain, Kiev, Czech, HCJB Ecuador -20.
1915 DLF, Germany (MW) dx T -20.
1920 Greece -1930.
1930 Finland -1945, Hungary, Yugo, Romania, Bulgaria, Canada -20, Iran -2030.
1935 Italy -1950.

20 Israel -2030, N Korea -2050, WCSN -22.
2005 Syria -2105.
2045 RBI -2130, Malta (7110) -2115.
2050 Vatican -2110.

21 Sweden dx T, Romania, Hungary, Japan -2130, Turkey -2150, Spain -22.
2130 HCJB dx S M, Bulgaria -22.

22 Finland, Belgium dx S -2225, China, Czech -2230, Canada dx S, Taiwan -2300.
2230 Israel, Albania, Poland dx T, Swiss -23, Vilnius, Malta -2330.
2245 RBI -2330, TWR -2335 $ 2345 T 2350.

23 Turkey -2350, Japan -24.
2305 Poland dx T -2355.

These timings are mainly intended for America, but are generally heard well in Europe. Try in particular the frequencies marked * in our shortwave list.

00 Israel, Sweden dx T, Canada -0130, Bulgaria -01, Cuba, Spain -02.
0030 Belgium, Hungary M-S -01, Netherlands dx Ŧ -0125.
0045 RBI -0130. Vatican 0050-0110.

01 Israel -0130, Deutsche Welle -0150, Canada S $ dx S -02, Czech -02.
0130 Austria, Hungary, Egypt -02.

02 Israel -0225, Swiss -0230, Romania, South Africa -03.
0230 Sweden -03, Hungary dx TWFS -0245.
0245 RBI -0330.

03 Deutsche Welle -0350, Czech, Japan, China -04, TWR Bonaire -0430.
0330 Netherlands dx Ŧ -0425.

04 Romania, Swiss -0430, RBI -0445, Bulgaria, Argentina -05.
0430 Austria -05.
0445 RBI -0530.

05 Israel -0515, Deutsche Welle -0550, Spain -06, Kuwait -08.

BBC World Service, Voice of America and Radio Moscow are excluded from our timings, as they are major stations which broadcast in English around the world 24 hours a day. They can generally be found throughout the day on the frequencies given, with the exception of Voice of America, for which the best time to listen is early morning and from 17 hours. Note: The brief BBC World Service "Waveguide" is broadcast $ 0750, M 0530, T 1115, Ŧ 0130.

Extra copies of the unique Dial-Search Maps are usually available, price 70p for the Europe map, 40p for the British Isles map, or £1.00 the two; plus 20p for postage and packing.

Readers interested in having Spot the Tune in full (notation of 80 signature tunes & interval signals: 4 pages) may write for a copy of Dial-Search 4th Edition, plus a partial up-date of Spot the Tune. (This copy includes the British Isles map but not the Europe map.) Price £1.50 post-free.

Remittances should be sent to George Wilcox, 9 Thurrock Close, Lower Willingdon, Eastbourne, East Sussex BN20 9NF.

Comments and up-date information are always welcome; enquiries too, preferably by telephone (03212.2419) otherwise please enclose return postage. We are not geared to advising on choice of receivers.

New readers may like to send us their name and full address, with one stamp or IRC, to be included in our mailing list for new editions.

48 SPOT THE TUNE

Getting to know the musical identity-card of a station is a great help to the regular listener, esp on SW. It may be a signature tune that starts a programme (at the hour or half-hour) or a repeated interval signal for the previous few minutes. At least one good cassette collection is available (see p8) but our feature "Spot the Tune" has been unrivalled for quick reference - and for accuracy - and the number allotted to each tune can be used in a listener's notes. The demand, however, is small, and for economy of space in this edition we give only a few which are esp helpful in MW listening. All are shown in the key of G (or G minor). See announcement on p47.

A SHORTWAVE SELECTION

This five-page list attempts to digest the output of the major broadcasting organisations. It should help the beginner to see the wood for the trees; and many readers find the layout useful for their own notes. At any given time of year, a good proportion of these frequencies will be audible in Europe at various times of night or day (the lower frequencies by day). Some will give long hours of listening. For proper up-to-date reference, other publications are essential (see p8). Consult also pages 7, 10 and 46-48.

* Starred frequencies: early mornings only.

	SPAIN PORTUGAL	VATICAN ITALY,MALTA	TURKEY GREECE	ISRAEL	AMERICAN, various (mainly religious prs)
MW:		1530 (V)			TWR 1467 [3]
		1557 (M)			
49				5915	[3] Monaco
metre		6110 (M)			[2] Ecuador
band		6185			# Sines
		6190			° Forli
41 mb					AWR 7125 °
		7250			AWR 7275 °
		7275 (I)			
		7290 (I)			
31 mb			9395 (G)		
			9425 (G)		TWR 9485 [3]
		9575 (I)			TWR * 9535
	9605 (P)	* 9605			HCJB 9610 [2]
	*9630	9645	9685		AWR 9670 #
	9740 (P)		9795	9855	WCSN 9840
25mb			11595 (G)	11585	
				11588	
	11740 (P)	11740		11605	HCJB * 11755
	11790	*11780			HCJB 11835 [2]
	*11880				WHRI 13760
	11920				WCSN 13770
19mb					HCJB *15155 & 15270 [2]
	15280				WYFR 15566
	15375				WCSN 15610
	15395				KUSW 15650
				15640	WWCR 15690
16mb				17575	WYFR 17845
				17590	
				17630	
				17685	
13mb					WYFR 21525

	BBC WORLD SERVICE	VOICE OF AMERICA	WEST GERMANY	AUSTRIA FRANCE	SWITZER-LAND	NETHERLANDS BELGIUM (BRT)
MW:	648, 198	1197, 792	1269 (DLF)			1512 (B)
75mb	3955	3980	all	3965 (F)	3985	
			Deutsche			
49mb			Welle:-	5945		5910 (B)
	5975	5995				5955
	6040	6040	* 6085	6045 (F)		* 6020
		6095	* 6145			
	6180 (C)		6160	6155	6165	
	6195		6170	6175 (F)	6190	
41mb	7150	7170		7135 (F)		
		7200	7225	7280 (F)		
	7325	7325				
31mb	9410					
	9580 (C)		* 9545		9535	* 9590
			9650			
	9750	9740	* 9670	9745 (F)	* 9725	9715
	9760	9760	* 9735	9805 (F)		
				* 9875	* 9885	9860
25mb				11670 (F)		11695 (B)
		11760	11785			* 11720
		11925	11945			
	12095			13730	* 12035	13700
19mb	15070		15105	15155 (F)		
	15260 (A)			15195 (F)		
	15310 (S)		15410			
	15400 (A)		15595			15560
16mb			17825	17620 (F)		17605
13mb			21600			
	(C) Cyprus relay					
	(A) Ascension Island					
11mb	relay					
	(S) Singapore relay					

	SWEDEN	FINLAND	USSR: Radio	USSR: Kiev,	ALBANIA	ROMANIA
	NORWAY		Moscow Wld	Peace & Pro-		
		LW 252;	Svce in Engl	gress, Vilnius		
MW:	1179	963,558	1143,1431,	1386 (P&P)	1395	
			1494	666 (V)		
75mb			4060	4795 (P&P)		
49mb	5980 (N)		5905 #	5905 (P&P)		* 5990
	6065		6000	6010,6090(K)		
			6030 #	6100 (V)		* 6155
		6120	6055	6145 (P&P)		
			6165	6165 (K)		
41mb			7150 #	7115 (K)	7120	
			7170 #			
	7265			7205 (P&P)	7205	
			7310	&7215, 7340,		
			7350, 7370,	7360, 7440.		
			7390	7400 (V)		
31mb			9450, 9515		9480	
	9590 (N)	9560	9685	9580 (P&P)	9500	
	9655	9640	9765	9610 (P&P)		9690
	* 9695		9795	9775 (P&P)		9750
			9860	9700 (V)		
			9870			
25mb	11705		11705			
		11755	11745			
		11850	11850	11980 (P&P)		11810
			12010			11940
19mb		15185		15180 (V)		
	15220 (N)		15260			15250
			15305			
			15345			15360
			15480	15455 (V)		
16mb			17810			17850
			17840			
13mb	21655		21660			
			# also			
11mb			Foreign			
			Service			

	HUNGARY	CZECH, BULGARIA	POLAND YUGOSLAVIA	EAST GERMANY	ARAB COUNTRIES	INDIA PAKISTAN AFGHANISTAN
MW:		1287 (Cz)	P: 1503, 1206, 738	1359		
49mb		5930	5955	5965		
		6055	5980 (Y)	6040 S S̸		
		6070 (B)		* 6080		6020 (A)
	6110			6115		
41mb		7115 (B)		7185		
		7155 (B)				
	7220		7215 (Y)	7260		7215 (A)
			7270			
			7285	7295		
		7345				7412 (I)
31mb					9022 Iran	
	9585	9505	9540		9560 Jordan	9550 (I)
			9620 (Y)	9665		9635 (A)
		9700 (B)			9705 Saudi	
	9835				9720 Saudi	
						9910 (I)
25mb		11660 (B)			11665 Kuwait	11570 (P)
		11720 (B)		* 11785	11785 Dubai	11620 (I)
	11910		11815		12085 Syria	
					13610 Kuwait	
					13660 Iraq	
19mb	15160	15160 (B)				
	15220	15330 (B)			15300 Dubai	
					15320 Dubai	
					15435 Dubai	
16mb					17775 Dubai	
		17825 (B)				
13mb					21605 Dubai	21575 (P)
11mb						

	SOUTH KOREA NORTH KOREA	CHINA TAIWAN	JAPAN	S AFRICA AUSTRALIA	CANADA	S AMERICA
75mb		3985, via Switzerland				
49mb			* 5960 via R Canada		* 5960	
				6035 (Aus)	5995 (D)	
					6050 (D)	
					6150	
	6575 (N Kor)					
41mb				7205 (Aus)	7155 (D)	
				7215 (Aus)	7235 (D)	
31mb	9325 (N Kor)					
	9345 (N Kor)					
				* 9615		9685 Cuba
		* 9770		9655 (Aus)	9740 (D)	
		9820			* 9755	
		9852 (T)			9760	
25mb		11500				
	11760 (N Kor)	* 11715		11805	11840	11710
		11805 (T)	11835 (G)	11900	11945	Argentina
				* 11935		
	12020 (V Nam)					
19mb	13670 (S Kor)					
	15010 (V Nam)					
		15345 (T)	15325		15325	15265
	15575 (S Kor)					Brazil,
16mb						R Bras
13mb			21690 (G)	21535		
			21700	21590		
					(D) =	
11mb			(G) Gabon	25790	Daventry	
			relay		relay	

USING THE DIAL-SEARCH MAPS

1. Readers obtaining their copy by mail-order from the publisher will find their home town already plotted on the map(s) by a pencil cross.

2. Other readers should look up their own latitude & longitude in the back of a good atlas, or enquire at their library. It is simple to plot this point on the Dial-Search map(s) as follows:-
(a) Mark the latitude (North) on the vertical line, also on the scale in the margin right or left; draw a horizontal line through the two marks.
(b) Mark your longitude (East or West) along the scales at the top & bottom of the map; join these marks with a vertical line.
The crossing point will mark your location (or QTH).
[Reminder: there are 60 minutes in one degree; examples:-]

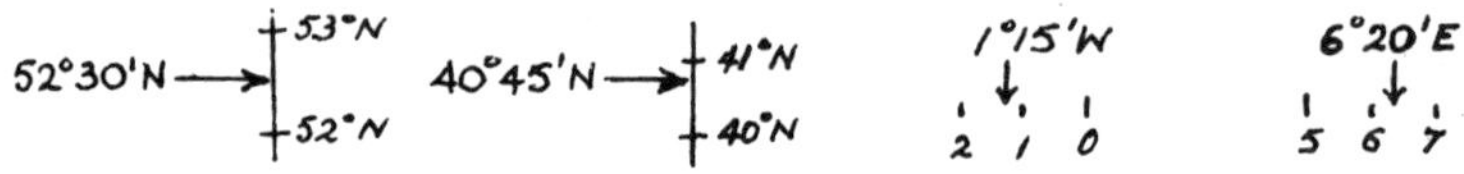

3. Any town can be plotted in the same way (a holiday location, a friend's home, a town you move to); these other crosses should be smaller, and make sure they are labelled.

4. The grid underprinted on the British Isles map also appears on the Europe map, making it quick to transfer a plotting from one to the other.

5. Line up a simple protractor on the pencil cross on the map(s), with 0° at the top, and 90° to the right. Mark off at 45° and 135° at least; rule pencil lines through the cross (dotted lines will do) across the whole map. Label the ends of these "spokes" either N, NE, E, SE etc, or 0, 45, 90, 135 etc, as you prefer. For portable radios (with their single ferrite aerial built in) there is no real need to go past 180°; just label South as 0 and continue through the same sequence back to North (e.g. SE and NW are both labelled 135). For greater precision, additional spokes will be needed; 20, 70, 110 & 160 are suggested; more than these would make the map less clear to use. See Fig.1 on page 8.

6. Find a transmitter location on the map and note its direction from you. If Prague, say, and its neighbouring towns are on or near the 90 line, that is where your set should face to receive them. In the MW list, in the column provided headed "bearing", write in "90" or "E" against the frequency (e.g. against 1287 kHz 233 m). Gradually your column will fill up and make successful tuning a rapid matter.

7. Any transmitter that we have omitted may be plotted as in 2 above.

8. The Europe map may be given an extension if desired to include more of Central Russia, the Arctic, or North Africa. Using 5 mm squared paper, the longitude scale is unchanged; the latitude scale south of 30° is virtually the same as from 30° to 35°N; north of 65° the degrees must go on "stretching" a little.

9. Those interested in reception from North America etc on MW will need to buy a large world map (only Mercator's Projection is suitable). For trans-polar bearings, or for great circle working (mainly shortwave), special maps are required; computer programmes also exist.

Most portable radios are astonishingly selective, simply because it is easy to set them at various angles. A heavier set, or a tuner built into a music centre, does not have this advantage, nor does a set with a fixed external aerial. (If a fixed site has to be chosen for the set, east or west facing is probably best value, in the U.K.)

Of course, many stations can be received over a wide angle, and a strong local station at almost any angle. But a portable that remains stationary is wasting at least half of its potential, certainly on the medium waveband.

At least find out where the North is in your home, so that you can turn the set to face approximately the direction indicated by our maps. But it is worth being more precise:-

Take a piece of plain card or stiff paper (A4 or larger) and halve its length by drawing a centre line. Place it with its long side along the front edge of a table or desk which is in a permanent position, say against a wall. Place a compass over the centre line and mark <u>True</u> North and South (accuracy pays; strictly, you should allow for the current magnetic deviation at your location). Draw in the North-South line and label it 0 at each end. Set a protractor on it and rule the "spokes" just as in 6 on page 21. Line the card up again with the table edge, and you have a compass base for your radio set.

The most practical radio is one with a turn-table fitted in the base (such as are still made by Roberts Radio, of West Molesey, Surrey). Stick a paper pointer exactly at the centre of the front, near the bottom. Measure the base of the set and draw its shape on the card, with its centre on the centre of the compass pattern. Get used to placing the set quickly in position on this shape, so that it will rotate uniformly.

The pointer will now give the bearing of reception.

Just a few models of set have the internal ferrite rod antenna placed across the side of the set instead of along the front. In this case follow the advice below but allow 90°.

Controlling our bearing gives us three advantages:-

1. When the set faces the direction given by the map, it will be receiving the maximum signal from the transmitter, giving the <u>loudest and clearest reception</u>. (In some locations, reflection from large buildings etc may distort the angle, in practice.)

2. An unwanted transmitter may be cancelled out ("nulled") by making the <u>end</u> of the set point in its direction, so that reception of this <u>interference</u> (QRM) is <u>at minimum</u> or even zero. Probably you will still be able to receive the other station, the one that interests you. This method may also deal with some electrical interference (QRN).

3. You can often <u>identify</u> an unknown station by nulling it out in this way and noting the bearing of the end of the set in that position. From the map it will be clear whether it is, say, Madrid or Vienna or Paris (all on 585 kHz 514 m) from this bearing. Turn the set back through 90° to the bearing you noted, and listen for announcements, musical clues etc (probably just before the hour) to confirm your identification.

These three effects can be startlingly increased by adding a loop aerial, but this must be at right angles to the set and both must turn together. A really old set will have only a loop aerial, wound inside the case; probably the side of the set will need to face the desired station. - Further advice on antennas can be found in free leaflets available from Radio Canada International and others, and from books, or you could consult your local dealer or radio club.

Whatever your installation, reception is better the higher your set (or aerial) is. A portable in an upstairs room, near a window if possible, is much more reliable than on the ground floor.

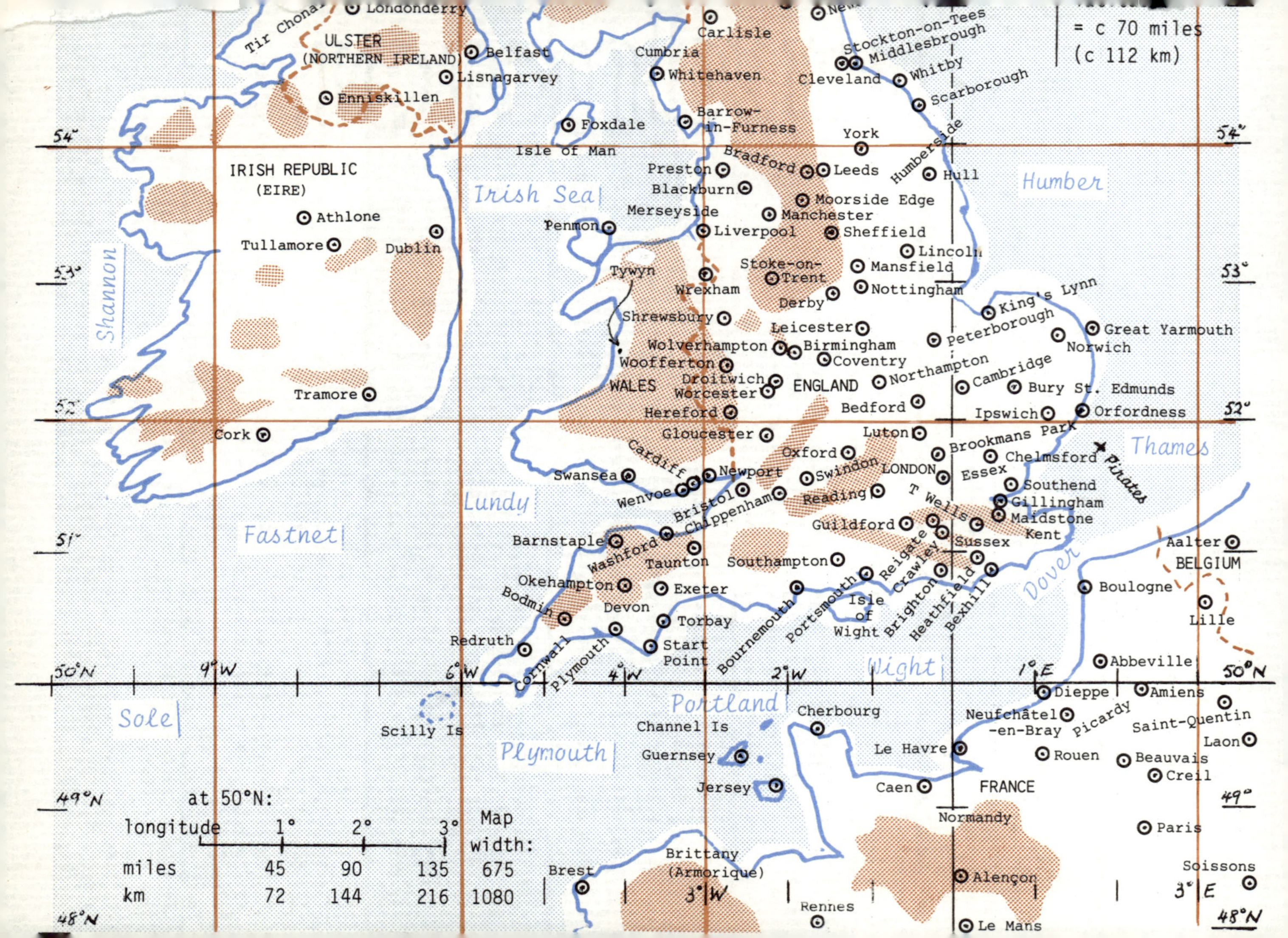

= c 70 miles
(c 112 km)
Londonderry
Tir Chonai
ULSTER
(NORTHERN IRELAND)
Belfast
Lisnagarvey
Enniskillen
IRISH REPUBLIC
(EIRE)
Athlone
Tullamore
Dublin
Tramore
Cork
Shannon
Fastnet
Sole
Lundy
Irish Sea
Plymouth
Portland
Wight
Dover
Thames
Humber
Isle of Man
Foxdale
Carlisle
Cumbria
Whitehaven
Barrow-in-Furness
Stockton-on-Tees
Middlesbrough
Cleveland
Whitby
Scarborough
York
Humberside
Hull
Preston
Bradford
Leeds
Blackburn
Moorside Edge
Merseyside
Manchester
Liverpool
Sheffield
Penmon
Lincoln
Stoke-on-Trent
Mansfield
Tywyn
Wrexham
Nottingham
Derby
King's Lynn
Shrewsbury
Leicester
Peterborough
Great Yarmouth
Norwich
Wolverhampton
Birmingham
Coventry
Woofferton
Droitwich
Worcester
WALES
ENGLAND
Northampton
Cambridge
Bury St. Edmunds
Bedford
Ipswich
Orfordness
Hereford
Gloucester
Luton
Brookmans Park
Chelmsford
Pirates
Cardiff
Oxford
Swansea
Newport
Swindon
LONDON
Essex
Southend
Wenvoe
Bristol
Chippenham
Reading
T Wells
Gillingham
Maidstone
Kent
Guildford
Sussex
Barnstaple
Washford
Taunton
Southampton
Reigate
Crawley
Okehampton
Exeter
Portsmouth
Isle of Wight
Brighton
Heathfield
Bexhill
Bodmin
Devon
Torbay
Bournemouth
Redruth
Cornwall
Plymouth
Start Point
Scilly Is
Channel Is
Guernsey
Jersey
Cherbourg
Le Havre
Caen
FRANCE
Normandy
Brittany
(Armorique)
Brest
Rennes
Alençon
Le Mans
Aalter
BELGIUM
Boulogne
Lille
Abbeville
Dieppe
Amiens
Neufchâtel-en-Bray
Picardy
Saint-Quentin
Laon
Rouen
Beauvais
Creil
Paris
Soissons
54°
53°
52°
51°
50°N
49°N
48°N
9°W
6°W
4°W
3°W
2°W
1°E
3°E
at 50°N:
longitude 1° 2° 3°
miles 45 90 135
km 72 144 216
Map width:
675
1080

The Dial – Search Map of the British Isles

Bearings constant. Projection: Mercator.

Scale at 60°N:

longitude	1°	2°	3°	Map width:
miles	35	70	105	525
km	56	112	168	840

Borders — Malin — Weather areas

High ground (c 200 m +)

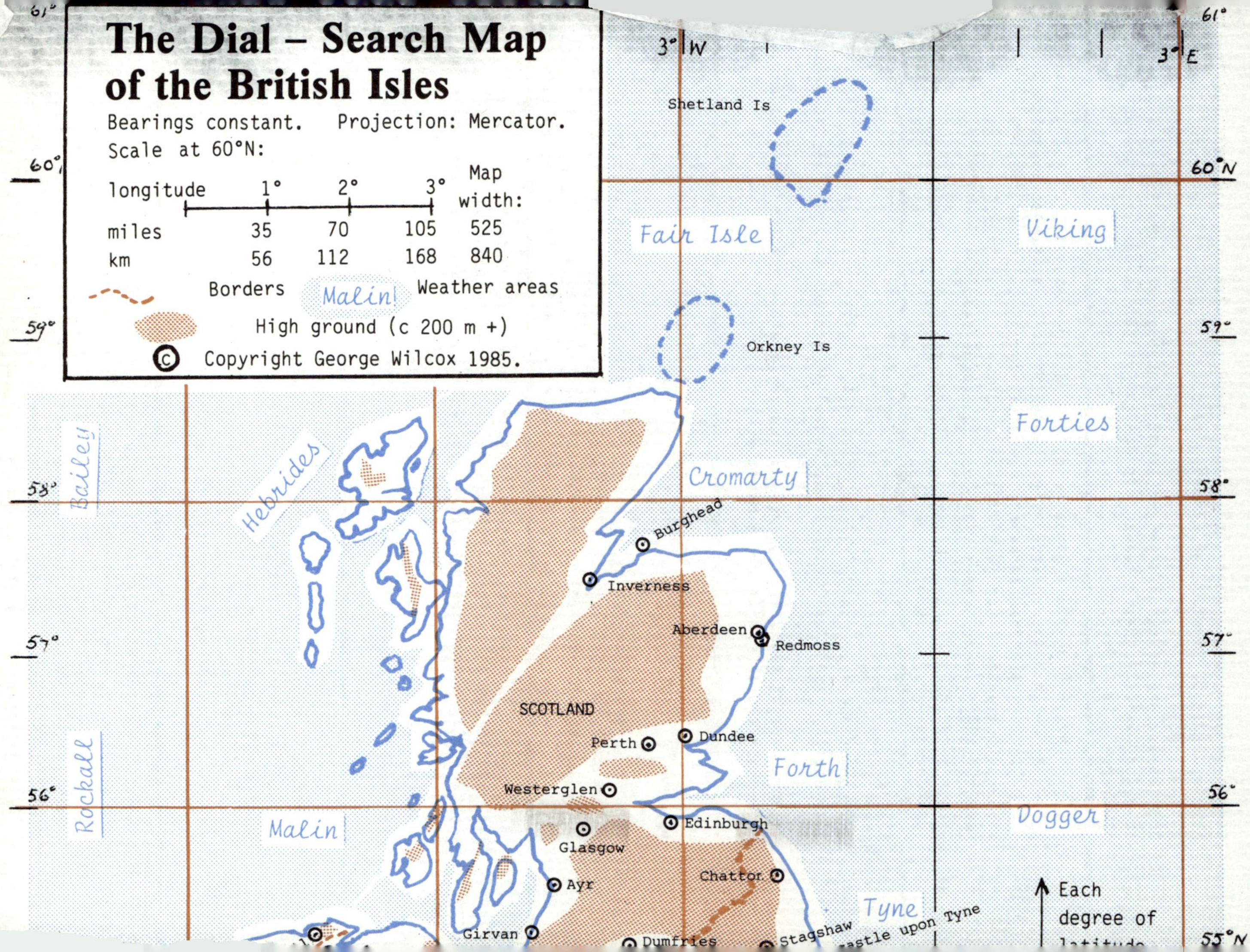

1. Readers obtaining their copy by mail-order from the publisher will find their home town already plotted on the map(s) by a cross in pencil.

2. Other readers should look up their own latitude and longitude in the back of a g atlas (or enquire at their library). It is simple to plot this point on the DIAL SEARCH map(s) as follows:-
   (a) Mark the latitude (North) on the vertical line, also on the scale in the ma right or left; draw a horizontal line through the two marks.
   (b) Mark your longitude (East or West) along the scales at the top and bottom o the map; join these marks with a vertical line.

   The crossing point will mark your location (or QTH).
   [Reminder: there are 60 minutes in one degree; examples:-]

3. Any town can be plotted in the same way (a holiday location, a friend's home, a town you move to); these other crosses should be smaller, and make sure they are labelled.

4. The grid underprinted on the U.K. map also appears on the Europe map, making it quic to transfer a plotting from one to the other.

5. It is recommended that the U.K. map be carefully removed from the staples and kep in an A4 plastic pocket. This also makes it easier to consult the U.K. list endir page 27.

6. Line up a simple protractor on the pencil cross on the map(s), with 0° at the top, ar 90° to the right. Mark off at 45° and 135° at least; rule pencil lines through the cross (dotted lines will do) across the whole map. Label the ends of these "spokes" either N, NE, E, SE etc, or 0, 45, 90, 135 etc, as you prefer. For portable radios (with their single ferrite aerial built in) there is no real need to go past 180°; ju label South as 0 and continue through the same sequence back to North (e.g. SE and N are both labelled 135). For greater precision, additional spokes will be needed; 70, 110 & 160 are suggested; more than these would make the map less clear to use.

7. If you are using a plastic pocket (see 5 above) this pattern of compass lines may be made on that, with a suitable waterproof pen, instead of on the U.K. map itself.

8. Find a transmitter location on the map and note its direction from you. If Prague, say, is on or near the 90 line, that is where your set should face to receive Prague In the MW list, in the column provided headed "bearing", write in "90" or "E" agai Prague (e.g. against 1233 kHz 243 m). Gradually your list will fill up and mak successful tuning a rapid matter.

9. Any transmitter that we have omitted may be plotted as in 2 above.

10. The Europe map may be given an extension if desired to include more of Central Russia the Arctic, or North Africa. Using 5 mm squared paper, the longitude scale is unchanged; the latitude scale south of 30°N is virtually the same as from 30° to 35 north of 65° the degrees must go on "stretching" a little. - Those interested in reception from North America etc on MW will need to buy a large world map, but onl Mercator's Projection is suitable. Special maps are required for trans-polar bea or for great circle working (mainly shortwave).